Zurich Lectures in Advanced Mathematics

Edited by

Mathematics in Zurich has a long and distinguished tradition, in which the writing of lecture notes volumes and research monographs play a prominent part. The *Zurich Lectures in Advanced Mathematics* series aims to make some of these publications better known to a wider audience. The series has three main constituents: lecture notes on advanced topics given by internationally renowned experts, graduate text books designed for the joint graduate program in Mathematics of the ETH and the University of Zurich, as well as contributions from researchers in residence at the mathematics research institute, FIM-ETH. Moderately priced, concise and lively in style, the volumes of this series will appeal to researchers and students alike, who seek an informed introduction to important areas of current research.

Previously published in this series:

Yakov B. Pesin, *Lectures on partial hyperbolicity and stable ergodicity*
Sun-Yung Alice Chang, *Non-linear Elliptic Equations in Conformal Geometry*
Sergei B. Kuksin, *Randomly forced nonlinear PDEs and statistical hydrodynamics in 2 space dimensions*
Pavel Etingof, *Calogero-Moser systems and representation theory*
Guus Balkema and Paul Embrechts, *High Risk Scenarios and Extremes – A geometric approach*
Demetrios Christodoulou, *Mathematical Problems of General Relativity I*
Camillo De Lellis, *Rectifiable Sets, Densities and Tangent Measures*
Paul Seidel, *Fukaya Categories and Picard–Lefschetz Theory*
Alexander H.W. *Schmitt, Geometric Invariant Theory and Decorated Principal Bundles*
Michael Farber, *Invitation to Topological Robotics*
Alexander Barvinok, *Integer Points in Polyhedra*

Published with the support of the Huber-Kudlich-Stiftung, Zürich

Preface

> For dealing with atoms involving many electrons the accurate quantum theory, involving a solution of the wave equation in many-dimensional space, is far too complicated to be practicable. One must therefore resort to approximate methods. (P.A.M. Dirac, 1930)

This book is not about Nature, but about tools for exploring Nature. It is written by a mathematician who does not himself apply these tools but who appreciates their uses, admires their shapes, analyses their function, and at times designs improved hand axes.

Quantum dynamics of molecules poses a variety of computational challenges that are presently at the forefront of research efforts in numerical analysis in a number of application areas: high-dimensional partial differential equations, multiple scales, highly oscillatory solutions, and geometric structures such as symplecticity and reversibility that are favourably preserved in discretizations. This text addresses such problems in quantum mechanics from the viewpoint of numerical analysis, illustrating them to a large extent on intermediate models between the Schrödinger equation of full many-body quantum dynamics and the Newtonian equations of classical molecular dynamics. The fruitful interplay between quantum dynamics and numerical analysis is emphasized: numerical algorithms originally developed for quantum dynamics sometimes find a much wider scope of application areas, and numerical analysis can contribute theoretical hindsight and novel algorithms to computational quantum dynamics.

This short book contains the extended lecture notes of a graduate course (Nachdiplomvorlesung) held at ETH Zürich in the winter semester 2007. It has five chapters: Chapter I gives a concise introduction to quantum mechanics and establishes basic notions and notation. Chapter II treats the Dirac–Frenkel time-dependent variational principle and reduced models for many-body quantum dynamics built on it. Numerical methods for linear time-dependent Schrödinger equations are studied in Chapter III, those for the non-linear equations of motion of reduced models in Chapter IV. Finally, Chapter V describes a novel numerical approach to the Schrödinger equation in semi-classical scaling, using Hagedorn wave packets. The selection of topics has been guided by the wish to include a wide variety of beautiful ideas and successful techniques, with no attempt at encyclopedic completeness.

In preparing the text, I have greatly benefitted from discussions with and comments by Erwan Faou, Michael Fey, Ludwig Gauckler, Vasile Gradinaru, Ernst Hairer, Marlis Hochbruck, Tobias Jahnke, Gabriele Kallert, Othmar Koch, Caroline Lasser, Katina Lorenz, Olavi Nevanlinna, Christoph Schwab, Stefan Teufel, and Gerhard Wanner. Kersten Schmidt was the very helpful assistant to my course who prepared all the figures, with the exception of those in Chapter V which were kindly provided by Vasile Gradinaru. My sincere thanks go to all of them, and also to the participants of my course in Zürich.

Tübingen, March 2008

Table of Contents

Chapter I.
Quantum vs. Classical Dynamics

In this introductory chapter we recapitulate basic elements of quantum mechanics, emphasizing relationships with classical mechanics and preparing for the later chapters in a reasonably self-contained way. There are, of course, many texts where this material is presented more extensively and from different viewpoints. To name but a few, we mention the gentle mathematically-minded introduction by Thaller (2000), the complementary but visually equally appealing physical approach by Brandt & Dahmen (2001), the substantial brief text by Gustafson & Sigal (2003) from the mathematical physics point of view and the outreaching book by Tannor (2007) with a time-dependent, chemical physics perspective. There are the monumental classic treatises by Messiah (1962) and Cohen-Tannoudji, Diu & Laloë (1977), and the historical milestones left behind by Dirac (1930) and von Neumann (1932).

I.1 A First Look

To enter the stage, we begin by formulating the equations of motion of one (or several) particles in classical and quantum mechanics. We consider a particle of mass m in a conservative force field, which is the negative gradient of a potential $V(x)$, $x \in \mathbb{R}^3$.

I.1.1 Classical Mechanics

In classical dynamics, the state of the particle at any time t is characterized by its *position* $q(t) \in \mathbb{R}^3$ and *momentum* $p(t) \in \mathbb{R}^3$. It changes in time according to the Newtonian equations of motion

$$m\ddot{q} = -\nabla V(q)\,, \qquad p = m\dot{q}\,,$$

where the dots denote differentiation with respect to time t ($\dot{} = d/dt$). This can equivalently be written as a first-order system of ordinary differential equations,

$$\begin{aligned} \dot{q} &= \frac{p}{m} \\ \dot{p} &= -\nabla V(q)\,. \end{aligned} \tag{1.1}$$

With the *Hamiltonian function*

$$H(q,p) = T(p) + V(q), \qquad T(p) = \frac{|p|^2}{2m}$$

(here $|p|^2 = p \cdot p$ is the squared Euclidean norm), which represents the total energy as the sum of the kinetic energy $T(p)$ and the potential energy $V(q)$, the differential equations become Hamilton's canonical equations of motion

$$\begin{aligned} \dot{q} &= \frac{\partial H}{\partial p}(q,p) \\ \dot{p} &= -\frac{\partial H}{\partial q}(q,p). \end{aligned} \tag{1.2}$$

The formalism extends in a straightforward way to a system of N particles of masses $m_1, \ldots, m_N$, with the position vector $q = (q_1, \ldots, q_N)^T \in \mathbb{R}^{3N}$ and the momentum vector $p = (p_1, \ldots, p_N)^T \in \mathbb{R}^{3N}$ collecting the positions and momenta of the particles. The kinetic energy is then given as the sum of the kinetic energies of the particles, $T(p) = \sum_{n=1}^{N} |p_n|^2/(2m_n)$, and the potential $V(q) = V(q_1, \ldots, q_N)$ depends on the positions of all the particles and characterizes their interaction. The potential might in addition also depend on time to describe phenomena in a time-varying environment. Adding one more particle has the consequence of adding six *dependent* variables $(q_{N+1}(t), p_{N+1}(t))$ to the system of ordinary differential equations. Computations with millions, even billions of particles are routinely done in classical molecular dynamics simulations.

I.1.2 Quantum Mechanics

In quantum mechanics, the state at time t is described by the complex-valued *wave function* $\psi(x,t)$, depending on $x \in \mathbb{R}^3$ in the case of a single particle. Motivated by de Broglie's hypothesis of a particle-wave duality of matter, Schrödinger (1926) postulated the evolution equation that has since been recognized as the fundamental law for describing non-relativistic particles in physics and chemistry:

$$i\hbar \frac{\partial \psi}{\partial t} = H\psi. \tag{1.3}$$

Here, $i = \sqrt{-1}$ is the imaginary unit, and $\hbar$ is Planck's constant which has the physical dimension of an action, that is, energy divided by frequency or momentum times length. Its value is $\hbar = 1.0546 \cdot 10^{-34}$ Joule·sec. The *Hamiltonian operator* H on the right-hand side is the sum

$$H = T + V \tag{1.4}$$

of the kinetic energy operator T and the potential V. Here,

$$T\psi = -\frac{\hbar^2}{2m} \Delta\psi \tag{1.5}$$

with the Laplacian $\Delta = \nabla \cdot \nabla$ (the divergence and gradient are with respect to the spatial variable x). With the *momentum operator* $p = -i\hbar\nabla$, the expression of the kinetic energy looks formally the same as in classical mechanics:

$$T\psi = \frac{p \cdot p}{2m}\psi .$$

On the other hand, the potential simply acts as a multiplication operator:

$$(V\psi)(x) = V(x)\psi(x).$$

The Schrödinger equation (1.3) is thus a partial differential equation of first order in time and second order in space.

The usual *statistical interpretation* of quantum mechanics, due to Born (1926), views $|\psi(\cdot,t)|^2$ as a probability density for the position of the particle: the probability of the particle to be located within a volume $\Omega \subset \mathbb{R}^3$ at time t, equals $\int_\Omega |\psi(x,t)|^2\, dx$. Moreover, the squared absolute value of the Fourier transform of the wave function is interpreted as the probability density for the momentum of the particle.

The formalism again extends directly to several particles. As in the classical case, the multi-particle Hamiltonian is constructed as the sum of the kinetic energies of the single particles and a potential accounting for external forces and interaction. The Hamiltonian operator now acts on a wave function $\psi(x_1,\dots,x_N,t)$ depending on the spatial coordinates corresponding to each of the N particles. Its squared absolute value represents the joint probability density of particles 1 to N to be at $(x_1,\dots,x_N)$ at time t. The multi-particle wave function is a high-dimensional object: adding one more particle yields another three *independent* variables! Computations with direct finite-difference discretizations of Schrödinger's equation are out of reach for more than two or three particles.

I.2 The Free Schrödinger Equation

In the absence of a potential, for $V = 0$, the Schrödinger equation (1.3) becomes

$$i\hbar\frac{\partial\psi}{\partial t}(x,t) = -\frac{\hbar^2}{2m}\Delta\psi(x,t)\,, \qquad x \in \mathbb{R}^d,\ t \in \mathbb{R}\,. \tag{2.1}$$

I.2.1 Dispersion Relation

Einstein's equation

$$E = \hbar\omega \tag{2.2}$$

relates the energy of emitted electrons to the frequency of incident light in the photoelectric effect, which is explained by light quanta showing the particle nature of light (Einstein 1905). It was hypothesized by de Broglie (1924) that particle-wave duality should exist also for matter, and the energy relation (2.2) should be basic also for matter waves. As we will see in a moment, Equation (2.1) can be understood as resulting from an effort to reconcile (2.2) with the classical expression for the energy of a free particle with mass m and momentum p,

$$E = \frac{|p|^2}{2m}\,, \tag{2.3}$$

for solutions of a linear evolution equation

$$\frac{\partial \psi}{\partial t} = P(\partial_x)\psi\,, \qquad x \in \mathbb{R}^d,\ t \in \mathbb{R}\,,$$

with some (possibly pseudo-) differential operator $P(\partial_x)$. A plane wave $e^{i(k\cdot x-\omega t)}$ with wave vector $k \in \mathbb{R}^d$ and angular frequency ω is a solution of this equation if ω satisfies the *dispersion relation*

$$\omega = \omega(k) = iP(ik)\,.$$

Clearly, knowing the dispersion relation is tantamount to knowing the evolution equation with operator $P(\partial_x)$.

In relating (2.2) and (2.3), it is assumed that the momentum should be

$$p = mv\,, \tag{2.4}$$

where the velocity is taken to be the *group velocity*

$$v = \frac{\partial \omega}{\partial k}\,, \tag{2.5}$$

which is the velocity of the envelope of a localized wave packet (Hamilton 1839, Rayleigh 1877; see also Sect. I.2.3 below) and thus represents the particle velocity. With the relations (2.4)–(2.5), the equality of the energies (2.2) and (2.3) becomes the condition $\hbar\omega = \frac{1}{2}m|\partial\omega/\partial k|^2$, which is satisfied for the dispersion relation of the free Schrödinger equation (2.1),

$$\hbar\omega = \frac{\hbar^2}{2m}\,|k|^2\,. \tag{2.6}$$

With (2.4)–(2.5), this further implies de Broglie's relation

$$p = \hbar k\,, \tag{2.7}$$

which together with (2.2) expresses the plane wave as $e^{i(k\cdot x-\omega t)} = e^{\frac{i}{\hbar}(p\cdot x-Et)}$. With (2.7), the equality of the energies (2.2) and (2.3) is just the dispersion relation (2.6) of the Schrödinger equation.

I.2.2 Solution by Fourier Transformation

We consider (2.1) together with the initial condition

$$\psi(x,0) = \psi_0(x)\,, \qquad x \in \mathbb{R}^d. \tag{2.8}$$

To concur with the interpretation of $|\psi_0|^2$ as a probability density, we assume that ψ_0 has unit L^2 norm:

$$\|\psi_0\|^2 = \int_{\mathbb{R}^d} |\psi_0(x)|^2 \, dx = 1 \,. \tag{2.9}$$

This initial-value problem is solved using Fourier transforms. We begin by recalling the necessary prerequisites; see, e.g., Katznelson (1976), Chap. VI, or Reed & Simon (1975), Chap. IX. For convenience, in the following we choose physical units such that

$$\hbar = 1.$$

Fourier Transform. Let $\mathcal{S}$ denote the Schwartz space of rapidly decaying smooth functions, that is, of arbitrarily differentiable complex-valued functions on $\mathbb{R}^d$ which, together with all their partial derivatives, decay faster than the inverse of any polynomial as $|x| \to \infty$. For a Schwartz function $\varphi \in \mathcal{S}$, the Fourier transform $\widehat{\varphi} = \mathcal{F}\varphi$ given by

$$\widehat{\varphi}(k) = \frac{1}{(2\pi)^{d/2}} \int_{\mathbb{R}^d} e^{-ik\cdot x} \, \varphi(x) \, dx \,, \qquad k \in \mathbb{R}^d, \tag{2.10}$$

is again a Schwartz function. There is the inversion formula

$$\varphi(x) = \frac{1}{(2\pi)^{d/2}} \int_{\mathbb{R}^d} e^{ik\cdot x} \, \widehat{\varphi}(k) \, dk \,, \qquad x \in \mathbb{R}^d, \tag{2.11}$$

and the Plancherel formula relating the L^2 norms of φ and $\widehat{\varphi}$,

$$\|\varphi\| = \|\widehat{\varphi}\| \,. \tag{2.12}$$

The Fourier transform changes partial derivatives into multiplication by the Fourier variable:

$$-\widehat{i\partial_j\varphi}(k) = k_j \widehat{\varphi}(k) \,. \tag{2.13}$$

The negative Laplacian is thus transformed into multiplication by the squared Euclidean norm $|k|^2 = k_1^2 + \cdots + k_d^2$:

$$-\widehat{\Delta\varphi}(k) = |k|^2 \widehat{\varphi}(k) \,. \tag{2.14}$$

By density or duality, the above formulas are extended to appropriate larger spaces of functions or distributions.

Solution via Fourier Transformation. Formally taking Fourier transforms with respect to the spatial variable x in (2.1) yields decoupled ordinary differential equations parametrized by the dual variable k:

$$i \frac{\partial \widehat{\psi}}{\partial t}(k,t) = \frac{|k|^2}{2m} \, \widehat{\psi}(k,t) \,, \qquad k \in \mathbb{R}^d,$$

which are solved by

$$\widehat{\psi}(k,t) = e^{-i\frac{|k|^2}{2m}t} \, \widehat{\psi}_0(k) \,. \tag{2.15}$$

Obviously, $|\widehat{\psi}(k,t)|^2 = |\psi_0(k)|^2$ for all k and t. We note that for initial data in the Schwartz space, $\psi_0 \in \mathcal{S}$, we have $\widehat{\psi}_0 \in \mathcal{S}$ and thus further $\widehat{\psi}(\cdot,t) \in \mathcal{S}$ for all real t. The function obtained by the inverse Fourier transform (2.11),

$$\psi(x,t) = \frac{1}{(2\pi)^{d/2}} \int_{\mathbb{R}^d} e^{i(k\cdot x - \frac{|k|^2}{2m}t)}\, \widehat{\psi}_0(k)\, dk\,, \tag{2.16}$$

is thus again a Schwartz function, and by the above transform rules, this function is verified to be a solution to (2.1) with (2.8). We have unit L^2 norm

$$\|\psi(\cdot,t)\|^2 = 1 \qquad \text{for all } t \tag{2.17}$$

by the Plancherel formula, by (2.15) and condition (2.9), so that $|\psi(\cdot,t)|^2$ remains a probability density for all times.

The Free-Evolution Operator. With the kinetic energy operator $T = -\frac{1}{2m}\Delta$, we use the notation

$$\psi(\cdot,t) = \psi(t) =: e^{-itT}\psi_0\,.$$

This defines the evolution operator

$$e^{-itT} : \mathcal{S} \to \mathcal{S}.$$

By (2.17) and because the Schwartz space $\mathcal{S}$ is dense in the Hilbert space L^2 of square-integrable functions, we can extend the operator to a norm-preserving operator

$$e^{-itT} : L^2 \to L^2,$$

and we consider $e^{-itT}\psi_0$ for arbitrary $\psi_0 \in L^2$ as a generalized solution to the free Schrödinger equation (2.1) with initial state (2.8).

I.2.3 Propagation of Heavy Wave Packets

We consider the free Schrödinger equation (2.1) with $\hbar = 1$, and as initial state a wave packet

$$\psi_0(x) = e^{ip\cdot x}\, a(x) \qquad \text{with} \quad a \in \mathcal{S},\ p \in \mathbb{R}^d. \tag{2.18}$$

We are particularly interested in p of large norm, so that a highly oscillatory complex exponential is modulated by the smooth, rapidly decaying function $a(x)$. The following relation will be shown for the solution of (2.1), uniformly in $p \in \mathbb{R}^d$ as the mass $m \to \infty$:

$$\psi(x,t) = e^{ip\cdot(x - \frac{p}{2m}t)}\, a\Big(x - \frac{p}{m}t\Big) + \mathcal{O}\Big(\frac{t}{m}\Big). \tag{2.19}$$

Here we note the *phase velocity* $p/(2m)$ in the argument of the exponential and the *group velocity* $v = p/m$ in the argument of a, and

$$|\psi(x,t)|^2 \approx |\psi_0(x - vt)|^2\,,$$

which describes uniform straight motion of the envelope with the group velocity v. In particular, the centre of the wave packet, $q(t) = \int_{\mathbb{R}^d} x\, |\psi(x,t)|^2\, dx$, moves according to

$$q(t) \approx q(0) + t\,\frac{p}{m}\,. \tag{2.20}$$

Heavy particles thus show approximately classical behaviour.

Proof of (2.19): We start from formula (2.16) for $\psi(x,t)$ and note that $\widehat{\psi}_0(k) = \widehat{a}(k-p)$. We decompose $|k|^2 = |k-p+p|^2 = |p|^2 + 2(k-p)\cdot p + |k-p|^2$ and substitute the integration variable k for $k-p$ to obtain

$$\psi(x,t) = e^{ip\cdot x}\, e^{-i\frac{|p|^2}{2m}t}\, \frac{1}{(2\pi)^{d/2}} \int_{\mathbb{R}^d} e^{ik\cdot x}\, e^{-ik\cdot\frac{p}{m}t}\, e^{-i\frac{|k|^2}{2m}t}\, \widehat{a}(k)\, dk\,.$$

With the relation $|e^{-i\frac{|k|^2}{2m}t} - 1| \le |k|^2 \frac{t}{m}$, we obtain with the inverse Fourier transform formula

$$\begin{aligned}
\psi(x,t) &= e^{ip\cdot x - i\frac{|p|^2}{2m}t}\, \frac{1}{(2\pi)^{d/2}} \int_{\mathbb{R}^d} e^{ik\cdot(x-\frac{p}{m}t)}\, \widehat{a}(k)\, dk + \mathcal{O}\Big(\frac{t}{m}\Big) \\
&= e^{ip\cdot x - i\frac{|p|^2}{2m}t}\, a\Big(x - \frac{p}{m}t\Big) + \mathcal{O}\Big(\frac{t}{m}\Big),
\end{aligned}$$

where the constant in the $\mathcal{O}$-symbol is $C = \int_{\mathbb{R}^d} |k|^2 |\widehat{a}(k)|\, dk$. □

I.3 The Schrödinger Equation with a Potential

We now turn to the Schrödinger equation (1.3) with a real-valued potential $V(x)$, $x \in \mathbb{R}^d$,

$$i\hbar\frac{\partial\psi}{\partial t} = -\frac{\hbar^2}{2m}\Delta\psi + V\psi. \tag{3.1}$$

For convenience we choose again units with $\hbar = 1$, as we will usually do when we treat mathematical rather than physical questions.

I.3.1 Self-Adjoint Operators and Existence of Dynamics

The existence of solutions to (3.1) rests on the theory of self-adjoint unbounded operators on a Hilbert space. Let us briefly recall the relevant concepts.

Let $\mathcal{H}$ be a complex Hilbert space with inner product $\langle\cdot\,|\,\cdot\rangle$, taken antilinear in its first and linear in its second argument. A linear operator $H : D(H) \to \mathcal{H}$, defined on a domain $D(H)$ dense in $\mathcal{H}$, is called *symmetric* if

$$\langle H\psi\,|\,\varphi\rangle = \langle\psi\,|\,H\varphi\rangle \qquad \forall \psi, \varphi \in D(H).$$

The operator is *self-adjoint* if for any $\varphi, \eta \in \mathcal{H}$ the relation

$$\langle H\psi \,|\, \varphi \rangle = \langle \psi \,|\, \eta \rangle \quad \forall \psi \in D(H) \qquad \text{implies} \qquad \varphi \in D(H) \text{ and } \eta = H\varphi .$$

Every self-adjoint operator is symmetric, but the converse is not true for unbounded operators. Every self-adjoint operator is *closed*: for any sequence (φ_n) in $D(H)$, the convergence $\varphi_n \to \varphi,\ H\varphi_n \to \eta$ implies $\varphi \in D(H)$ and $\eta = H\varphi$.

An operator U on $\mathcal{H}$ is *unitary* if it preserves the inner product:

$$\langle U\psi \,|\, U\varphi \rangle = \langle \psi \,|\, \varphi \rangle \qquad \forall\, \psi, \varphi \in \mathcal{H},$$

or equivalently, if it preserves the norm: $\|U\varphi\| = \|\varphi\|$ for all $\varphi \in \mathcal{H}$. As the following theorem states, for self-adjoint operators H the abstract Schrödinger equation

$$i\frac{d\psi}{dt} = H\psi \tag{3.2}$$

has a unitary evolution.

Theorem 3.1 (Existence of Dynamics). *Assume that H is a self-adjoint operator on a Hilbert space $\mathcal{H}$. Then, there is a unique family of unitary operators $e^{-itH},\ t \in \mathbb{R}$, with the following properties:*

1. *The operators e^{-itH} have the group property:*

$$e^{-i(t+s)H} = e^{-itH}\, e^{-isH} \qquad \textit{for all } s, t \in \mathbb{R}.$$

2. *The mapping $t \mapsto e^{-itH}$ is strongly continuous: for every $\psi_0 \in \mathcal{H}$,*

$$e^{-itH}\psi_0 \to \psi_0 \quad \textit{in the } \mathcal{H}\textit{-norm as } t \to 0.$$

3. *Equation (3.2) with initial value $\psi_0 \in D(H)$ has the solution $\psi(t) = e^{-itH}\psi_0$:*

$$i\frac{d}{dt}e^{-itH}\psi_0 = He^{-itH}\psi_0 ,$$

where the expressions on both sides of the equality sign indeed exist.

Theorem 3.1 can be proved by first noting that it holds for bounded operators, then by approximating H by a sequence of symmetric bounded operators H_n and carefully passing to the limit in the exponentials e^{-itH_n}; see Gustafson & Sigal (2003), Chap. 2. Another proof is based on the spectral theory of self-adjoint operators as developed by von Neumann and put to good use in his mathematical foundations of quantum mechanics (von Neumann, 1932). Based on von Neumann's spectral theory, Theorem 3.1 was given by Stone (1932) who also proves an interesting converse: if $U(t),\ t \in \mathbb{R}$, is a strongly continuous group of unitary operators, then $U(t) = e^{-itH}$ for some self-adjoint operator H.

I.3.2 Potentials Giving Self-Adjoint Operators

In applying Theorem 3.1 to the Schrödinger equation (3.1) on the Hilbert space $\mathcal{H} = L^2(\mathbb{R}^d)$ of square-integrable functions, we need conditions that ensure that the Hamiltonian $H = T + V$ is a self-adjoint operator. While symmetry is easily obtained, showing self-adjointness can be quite subtle.

First we remark that $T = -\frac{1}{2m}\Delta$ is self-adjoint with the domain $D(T) = H^2(\mathbb{R}^d)$, the Sobolev space of functions which together with their generalized partial derivatives up to order 2 are square integrable. (This is shown using Fourier transforms.)

Knowing that T is self-adjoint, what can we say about $T+V$? The following very useful perturbation result is known as the *Kato–Rellich theorem*, see Kato (1980), Sect. V.4.1, Theorem 4.3:

Let T be a self-adjoint operator on a Hilbert space, and V a symmetric operator bounded by $\|V\psi\| \le a\|\psi\| + b\|T\psi\|$ for all $\psi \in D(T)$, with $b < 1$. Then, $H = T + V$ is self-adjoint with domain $D(H) = D(T)$.

In particular, for $T = -\frac{1}{2m}\Delta$ a bounded potential always gives a self-adjoint Hamiltonian $H = T + V$ with domain $H^2(\mathbb{R}^d)$. A simple criterion that follows from the above result with the Sobolev inequality on $\mathbb{R}^3$, is the following (Kato 1980, Sect. V.5.3): Assume

$$V = V_\infty + V_2 \qquad \text{with} \quad V_\infty \in L^\infty(\mathbb{R}^3),\; V_2 \in L^2(\mathbb{R}^3).$$

Then, $T + V$ is self-adjoint with domain $D(H) = H^2(\mathbb{R}^3)$. For example, this applies to the *Coulomb potential* $V(x) = |x|^{-1}$.

An enlightening discussion and a variety of results on the self-adjointness of Schrödinger operators are given in Chapter X of Reed & Simon (1975). Remarkably, self-adjoint extensions always exist for a potential bounded from below (*ibid.*, p. 177), but they need not be unique, and different extensions can correspond to different physics (*ibid.*, p. 145). A unique self-adjoint extension is known to exist for every non-negative continuous *confining* potential, that is, satisfying $V(x) \to \infty$ as $|x| \to \infty$; see Hislop & Sigal (1996).

Later in this text, we will not pay much attention to the subtleties of self-adjointness and, in cases of possible doubt, we simply assume that the potential is such that $H = T + V$ yields a well-defined self-adjoint operator on $L^2(\mathbb{R}^d)$.

I.3.3 Lie–Trotter Product Formula

We have already constructed the free-particle evolution operator e^{-itT} by Fourier transformation, and for the potential we simply have $\big(e^{-itV}\psi\big)(x) = e^{-itV(x)}\psi(x)$. We do *not* have $e^{-i(T+V)} = e^{-itT}\, e^{-itV}$, but there is the following result due to Trotter (1959), whose finite-dimensional version is credited to Lie. See also Reed & Simon (1972), Theorem VIII.30, for precisely this version and a short proof.

Theorem 3.2. *Suppose that T, V, and $H = T+V$ are self-adjoint operators on a Hilbert space $\mathcal{H}$. Then, for every $t \in \mathbb{R}$ and $\varphi \in \mathcal{H}$,*

$$e^{-itH}\varphi = \lim_{n\to\infty} \Big(e^{-itT/n}\, e^{-itV/n}\Big)^n \varphi\,.$$

In view of the strong continuity of e^{-itV}, an equivalent statement is

$$e^{-itH}\varphi = \lim_{n\to\infty}\left(e^{-itV/(2n)}\, e^{-itT/n}\, e^{-itV/(2n)}\right)^n \varphi\,.$$

We will encounter the short-time approximation (over a small time step Δt)

$$e^{-i\Delta t H} \approx e^{-i\Delta t V/2}\, e^{-i\Delta t T}\, e^{-i\Delta t V/2}\,, \tag{3.3}$$

known as *symmetric Trotter splitting* or *Strang splitting*, repeatedly in this text, in various versions and disguises. This is one of the most widely used approximations to the evolution operator in computations.

Relationship with the Störmer–Verlet Method for Classical Mechanics. Consider now a wave packet as in Sect. I.2.3, $\psi_0(x) = e^{ip\cdot x}a(x)$, where we think of $a(x)$ as being localized near $x = q$. We consider the Taylor expansion of the potential $V(x)$ at q,

$$V(x) = V(q) + \nabla V(q)\cdot(x-q) + Q(x,q)$$

with the quadratic remainder term Q, so that we have

$$e^{-i\frac{\Delta t}{2}V(x)}\psi_0(x) = e^{i(p\cdot q - \frac{\Delta t}{2}V(q))}\, e^{i(p-\frac{\Delta t}{2}\nabla V(q))\cdot(x-q)}\, e^{-i\frac{\Delta t}{2}Q(x,q)}\, a(x)\,. \tag{3.4}$$

Here the first exponential on the right-hand side carries a phase which is modified by $-\frac{\Delta t}{2}V(q)$ over the half-step $\frac{\Delta t}{2}$. More interesting to us, in the second exponential the *momentum* p is shifted to $p - \frac{\Delta t}{2}\nabla V(q)$. We recall that in (2.20) we had a shift from *position* q to $q + \Delta t\, p/m$ for the centre of the wave packet propagated by the free evolution operator $e^{-i\Delta t T}$ in the situation of a large mass m. Combining these formulas for changing momenta and positions as they appear from the composition $e^{-i\Delta t V/2}\, e^{-i\Delta t T}\, e^{-i\Delta t V/2}$ of (3.3), we arrive at the following scheme: starting from q^0, p^0, set

$$\begin{aligned} p^{1/2} &= p^0 - \frac{1}{2}\Delta t\, \nabla V(q^0) \\ q^1 &= q^0 + \Delta t\, \frac{p^{1/2}}{m} \\ p^1 &= p^{1/2} - \frac{1}{2}\Delta t\, \nabla V(q^1)\,. \end{aligned} \tag{3.5}$$

This is the *Störmer–Verlet method* for the numerical solution of the Newtonian equations of motion (1.1), which is by far the most widely used numerical integration method in classical molecular dynamics. See Hairer, Lubich & Wanner (2003) for a discussion of this basic numerical method and its remarkable properties. We further note from (2.19) and (3.4) that the overall phase (the term in the exponential that is independent of x) is modified to

$$\phi^1 = \phi^0 + \Delta t\, \frac{|p^{1/2}|^2}{2m} - \frac{1}{2}\Delta t\,\bigl(V(q^0) + V(q^1)\bigr),$$

where the increment is a quadrature formula approximation to the classical action integral $\int_0^{\Delta t}\bigl(\frac{|p(t)|^2}{2m} - V(q(t))\bigr)\,dt$ along the solution $(q(t), p(t))$ of the classical equations of motion (1.1). We will explore relationships between (3.3) and (3.5) in more depth in Chapter V.

I.4 Averages, Commutators, Uncertainty

We consider again the Schrödinger equation (3.1) with the Hamiltonian $H = T + V$ and look at spatial averages of position, momentum, etc. along the wave function.

I.4.1 Observables and Averages

With the jth position coordinate as multiplication operator, $(q_j\psi)(x) = x_j\psi(x)$, and a function ψ of unit L^2 norm with $q_j\psi \in L^2$, we associate

$$\langle \psi \,|\, q_j\psi \rangle = \int_{\mathbb{R}^d} x_j \, |\psi(x)|^2 \, dx \,,$$

which represents the jth component of the *position average* of the state ψ, that is, the expectation value of the jth component of the position with respect to the probability density $|\psi|^2$.

With the jth component of the momentum operator, $p_j = -i\hbar\, \partial/\partial x_j$, we form (for $\psi \in D(p_j)$ and of unit L^2 norm)

$$\langle \psi \,|\, p_j\psi \rangle = \int_{\mathbb{R}^d} \overline{\psi}(x) \Big(-i\hbar \frac{\partial \psi}{\partial x_j} \Big) \, dx = \int_{\mathbb{R}^d} \hbar k_j \, |\widehat{\psi}(k)|^2 \, dk \,,$$

which is the jth component of the *momentum average* of the state ψ (recall de Broglie's relation (2.7): $p_j = \hbar k_j$). Similarly, we can consider the *total energy* $\langle \psi \,|\, H\psi \rangle$. It is such averages that can be observed experimentally.

Noting that q_j, p_j, H are self-adjoint operators on L^2, more generally we call any self-adjoint operator $A : D(A) \to L^2$ an *observable*. Its *average in the state* ψ (ψ of unit L^2 norm and $\psi \in D(A)$) is written, in varying notations,

$$\langle A \rangle = \langle A \rangle_\psi = \langle \psi \,|\, A \,|\, \psi \rangle = \langle \psi \,|\, A\psi \rangle \,. \tag{4.1}$$

I.4.2 Heisenberg Picture and Ehrenfest Theorem

Evolution of Averages and the Heisenberg Picture. We now study how the average $\langle A \rangle(t) = \langle A \rangle_{\psi(t)}$ of an observable A changes in time along a solution $\psi(t) = \psi(\cdot, t)$ of the Schrödinger equation (1.3). Since $\psi(t) = e^{-itH/\hbar}\psi_0$, we have

$$\langle A \rangle_{\psi(t)} = \langle A(t) \rangle_{\psi_0} \qquad \text{with} \quad A(t) = e^{itH/\hbar} A e^{-itH/\hbar} \,. \tag{4.2}$$

The operator $A(t)$ is said to give the *Heisenberg picture* of the evolution of the observable (after Heisenberg, 1925), as opposed to the Schrödinger picture working with wave functions. For a fixed initial state ψ_0, Eq. (4.2) can be written more briefly as

$$\langle A \rangle(t) = \langle A(t) \rangle \,. \tag{4.3}$$

Heisenberg Equation. A formal calculation, using the familiar relations $\frac{d}{dt}e^{-itH/\hbar} = \frac{1}{i\hbar}He^{-itH/\hbar} = e^{-itH/\hbar}\frac{1}{i\hbar}H$, yields

$$\frac{dA}{dt}(t) = \frac{1}{i\hbar}e^{itH/\hbar}\,(-HA + AH)\,e^{-itH/\hbar}$$

and hence, with the *commutator* $[A, H] = AH - HA$, we have the *Heisenberg equation*

$$\frac{dA}{dt}(t) = \frac{1}{i\hbar}\,[A(t), H]\,. \tag{4.4}$$

Remark. Some care is needed in giving a precise meaning to the commutator of unbounded self-adjoint operators, which in general need not exist. We note, however, that for initial states ψ_0 in a domain D which A maps into $D(H)$ and H maps into $D(A)$, the averages of both sides of (4.4) are well-defined and are indeed equal.

Energy Conservation. Since H commutes with itself, we obtain from (4.3) and (4.4) that the total energy is conserved along every solution of the Schrödinger equation:

$$\frac{d}{dt}\langle H\rangle(t) = 0. \tag{4.5}$$

Formal Analogy with Classical Mechanics. The Heisenberg equation (4.4) shows a close analogy to the corresponding situation in classical mechanics: a real-valued function $F(q,p)$ along a solution $(q(t), p(t))$ of the Hamiltonian equations (1.2) changes according to

$$\frac{d}{dt}F(q(t), p(t)) = \{F, H\}(q(t), p(t))$$

with the *Poisson bracket*

$$\{F, G\} = \sum_{j=1}^{d}\Big(\frac{\partial F}{\partial q_j}\frac{\partial G}{\partial p_j} - \frac{\partial F}{\partial p_j}\frac{\partial G}{\partial q_j}\Big),$$

as is seen by the chain rule and using (1.2). Formally, one bracket thus replaces the other in going from classical to quantum mechanics.

We now consider the Heisenberg equations (4.4) for the position and momentum operators, with components q_j and p_j: the corresponding time-dependent operators $q_j(t) = e^{itH/\hbar}q_je^{-itH/\hbar}$ and $p_j(t) = e^{itH/\hbar}p_je^{-itH/\hbar}$ satisfy (again with $\dot{} = d/dt$)

$$\begin{aligned}\dot{q}_j(t) &= \frac{1}{i\hbar}\,[q_j(t), H]\\ \dot{p}_j(t) &= \frac{1}{i\hbar}\,[p_j(t), H]\,.\end{aligned}$$

We note that $[q_j(t), H] = e^{itH/\hbar}[q_j, H]e^{-itH/\hbar}$ and similarly for $p_j(t)$. For a Hamiltonian $H = T + V$ with kinetic energy $T = -\frac{\hbar^2}{2m}\Delta$ and with a potential $V(x)$ acting as a multiplication operator, we calculate

$$\frac{1}{i\hbar}[q_j, H]\psi = \frac{1}{i\hbar}[q_j, T]\psi = -\frac{1}{i\hbar}\sum_{\ell=1}^{d}\frac{\hbar^2}{2m}[q_j, \frac{\partial^2}{\partial x_\ell^2}]\psi = -\frac{i\hbar}{m}\frac{\partial\psi}{\partial x_j} = \frac{p_j}{m}\psi$$
$$\frac{1}{i\hbar}[p_j, H]\psi = \frac{1}{i\hbar}[p_j, V]\psi = -\frac{\partial}{\partial x_j}(V\psi) + V\frac{\partial\psi}{\partial x_j} = -\frac{\partial V}{\partial x_j}\psi\,.$$

This gives Heisenberg equations that look like the classical equations of motion (1.1):

$$\begin{aligned} \dot{q}(t) &= \frac{p(t)}{m} \\ \dot{p}(t) &= -\nabla V(t) \end{aligned} \tag{4.6}$$

with $\nabla V(t) = e^{itH/\hbar}\,\nabla V\, e^{-itH/\hbar}$.

Ehrenfest Theorem. When we take averages $\langle\cdot\rangle$ on both sides of (4.6) according to (4.3), then we obtain the result by Ehrenfest (1927) that the position and momentum averages evolve by Newton-like equations:

$$\begin{aligned} \frac{d}{dt}\langle q\rangle &= \frac{\langle p\rangle}{m} \\ \frac{d}{dt}\langle p\rangle &= -\langle\nabla V\rangle\,. \end{aligned} \tag{4.7}$$

It should be noted, however, that in general

$$\langle\nabla V\rangle \neq \nabla V\big(\langle q\rangle\big)\,,$$

unless the potential is quadratic.

I.4.3 Heisenberg Uncertainty Relation

Still in analogy with classical mechanics, position and momentum are *canonically conjugate* observables, which here means that they satisfy (with Kronecker's delta)

$$\frac{1}{i\hbar}[q_j, p_k] = \delta_{jk}\,, \tag{4.8}$$

as is readily verified by a direct calculation of the commutator similar to the one given above. This has an important consequence to which we turn next. We define the *standard deviation* or *uncertainty width* of an observable A in a state ψ as

$$\Delta A = \big\langle (A - \langle A\rangle)^2\big\rangle^{1/2}\,, \tag{4.9}$$

where the average is taken with respect to the given state ψ.

Theorem 4.1 (Heisenberg Uncertainty Relation). *The standard deviations of the position and momentum operators satisfy the inequality*

$$\Delta q_j \, \Delta p_j \geq \frac{\hbar}{2} \,. \tag{4.10}$$

According to Heisenberg (1927), this now world-famous inequality is interpreted as saying that it is impossible to know at the same time both the position and momentum of an object with arbitrarily small uncertainty .

Proof. The result follows from (4.8) and from the *Robertson-Schrödinger relation* which states that for any observables A and B,

$$\Delta A \, \Delta B \geq \frac{1}{2} \Big| \big\langle [A, B] \big\rangle \Big| \,. \tag{4.11}$$

This is obtained with the Cauchy-Schwarz inequality and the identity

$$-2 \,\mathrm{Im}\, \langle A\psi \,|\, B\psi \rangle = \langle \psi \,|\, i[A, B]\psi \rangle$$

as follows (we may assume $\langle A \rangle = \langle B \rangle = 0$ for ease of notation):

$$\Delta A \, \Delta B = \|A\psi\| \cdot \|B\psi\| \geq \big| \langle A\psi \,|\, B\psi \rangle \big| \geq \big| \mathrm{Im}\, \langle A\psi \,|\, B\psi \rangle \big| = \frac{1}{2} \big| \langle i[A, B] \rangle \big| \,. \qquad \square$$

I.5 Many-Body Systems

I.5.1 Distinguishable Particles

Consider first N independent free particles, without any interaction, numbered from $n = 1, \ldots, N$. The probability density at time t for particle n to be at position x_n is $|\psi_n(x_n, t)|^2$, the square of the absolute value of the wave function. Since the particles are assumed independent, the joint probability density for particle 1 at $x_1, \ldots,$ particle N at x_N is the product $\prod_{n=1}^N |\psi_n(x_n, t)|^2$, which is the squared absolute value of the product wave function $\psi(x_1, \ldots, x_N, t) = \prod_{n=1}^N \psi_n(x_n, t)$ that solves the $3N$-dimensional free Schrödinger equation

$$i\hbar \frac{\partial \psi}{\partial t} = T\psi \quad \text{with} \quad T = \sum_{n=1}^{N} T_n \,, \quad T_n = -\frac{\hbar^2}{2m_n} \Delta_n \,,$$

where Δ_n is the Laplacian with respect to the variable x_n. Similarly, if each particle is subjected to an external potential $V_n(x_n)$, then the product wave function solves a Schrödinger equation with a potential that is the sum of the single-particle potentials. With particles interacting via a potential $V(x_1, \ldots, x_N)$, however, the solution of the multi-particle Schrödinger equation

$$i\hbar \frac{\partial \psi}{\partial t} = H\psi\,, \qquad H = T + V\,, \qquad \psi = \psi(x_1, \dots, x_N, t)\,,$$

is in general no longer in product form. As a rough *approximation* to the high-dimensional wave function we might still look for a function in product form — an old idea realized in the time-dependent *Hartree method* discussed in Chapter II.

I.5.2 Indistinguishable Particles

When particles cannot be distinguished in their physical properties, such as mass, charge, or spin, then the average of any observable is required to remain unchanged under an exchange of the particles. More formally, for two identical particles 1 and 2, denote by $R\psi(x_1, x_2, t) = \psi(x_2, x_1, t)$ the wave function for exchanged particles. It is then required that for every observable A and at every time t,

$$\langle R\psi \,|\, A \,|\, R\psi \rangle = \langle \psi \,|\, A \,|\, \psi \rangle\,. \tag{5.1}$$

To see the implications of this condition, consider the decomposition of the wave function into its symmetric and antisymmetric parts: $\psi = \psi_+ + \psi_-$ with $R\psi_+ = \psi_+$ and $R\psi_- = -\psi_-$. Condition (5.1) then yields Re $\langle \psi_+ \,|\, A \,|\, \psi_- \rangle = 0$ for all observables A, which turns out to imply that either $\psi_- = 0$ or $\psi_+ = 0$.

We are thus left with two possibilities, symmetry or antisymmetry:

$$\psi(x_2, x_1, t) = \psi(x_1, x_2, t) \qquad \text{(bosons)} \qquad \text{or} \tag{5.2}$$

$$\psi(x_2, x_1, t) = -\psi(x_1, x_2, t) \qquad \text{(fermions).} \tag{5.3}$$

Remarkably, for one kind of physical particle, always one and the same of the two cases is realized. The two situations lead to very different physical behaviour. It is the antisymmetry (5.3) that is known to hold for electrons, protons and neutrons: these are *fermions*. They obey the *Pauli exclusion principle* (Pauli 1925) which postulates that like particles cannot simultaneously be in the same quantum state. Note that (5.3) implies

$$\psi(x, x, t) = 0\,,$$

so that two identical fermions cannot be at the same position at the same time.

A product state does not have the antisymmetric behaviour (5.3) but it can be antisymmetrized: with two indistinguishable particles,

$$\psi(x_1, x_2, t) = \frac{1}{\sqrt{2}} \Big(\varphi_1(x_1, t)\varphi_2(x_2, t) - \varphi_1(x_2, t)\varphi_2(x_1, t) \Big)$$

has the required antisymmetry (and vanishes if $\varphi_1(\cdot, t) = \varphi_2(\cdot, t)$, in accordance with the Pauli principle), and so does the *Slater determinant*

$$\psi(x_1, \dots, x_N, t) = \frac{1}{\sqrt{N!}} \det \big(\varphi_j(x_n, t) \big)_{j,n=1}^{N} \tag{5.4}$$

in the case of N identical particles. Approximation of the electronic wave function by Slater determinants of orthogonal orbitals (i.e., single-electron wave functions) φ_j is done in the time-dependent *Hartree–Fock method*, see Chapter II.

It is also of interest to see what is the effect of ignoring antisymmetry in the approximation of the wave function of well-separated identical fermionic particles. Suppose that $\phi(x_1, x_2, t)$ is a solution of the time-dependent Schrödinger equation which is essentially localized near $(\langle x_1\rangle(t), \langle x_2\rangle(t))$ but which is not antisymmetric. As long as $\langle x_1\rangle(t)$ and $\langle x_2\rangle(t)$ remain clearly separated (well beyond the widths $\Delta x_1(t)$ and $\Delta x_2(t)$), the antisymmetrization $\phi(x_1, x_2, t) - \phi(x_2, x_1, t)$ does not deviate substantially from $\phi(x_1, x_2, t)$ in a neighbourhood of $(\langle x_1\rangle(t), \langle x_2\rangle(t))$, so that the particles can be considered to be distinguishable by their well-separated positions. This observation often justifies ignoring antisymmetry in the treatment of identical *nuclei* of a molecule, for which the above localization and separation condition is usually met in chemistry. On the other hand, for the less localized *electrons* a careful treatment of antisymmetry is essential.

I.5.3 The Molecular Hamiltonian

For a molecule, the Hamiltonian is the sum of the kinetic energy of the nuclei and the electrons, and the potential is the sum of the Coulomb interactions of each pair of particles:

$$H_{\mathrm{mol}} = T + V \quad \text{with} \quad T = T_N + T_e \quad \text{and} \quad V = V_{NN} + V_{Ne} + V_{ee}\,. \tag{5.5}$$

For N nuclei of masses M_n and electric charges $Z_n e$, with position coordinates $x_n \in \mathbb{R}^3$, and L electrons of mass m and charge $-e$, with coordinates $y_\ell \in \mathbb{R}^3$, the respective kinetic energy operators are

$$T_N = -\sum_{n=1}^{N} \frac{\hbar^2}{2M_n}\,\Delta_{x_n} \qquad T_e = -\sum_{\ell=1}^{L} \frac{\hbar^2}{2m}\,\Delta_{y_\ell}\,,$$

and the potential is the sum of the nucleus-nucleus, nucleus–electron and electron–electron interactions given by

$$V_{NN}(x) = \sum_{1\le k<n\le N} \frac{Z_k Z_n e^2}{|x_k - x_n|}\,, \qquad V_{Ne}(x,y) = -\sum_{\ell=1}^{L}\sum_{n=1}^{N} \frac{Z_n e^2}{|y_\ell - x_n|}\,,$$

$$V_{ee}(y) = \sum_{1\le j<\ell\le L} \frac{e^2}{|y_j - y_\ell|}\,.$$

It is often convenient to choose *atomic units* where $\hbar = 1$, the elementary charge $e = 1$, the mass of the electron $m = 1$, and where the length unit is chosen such that the Bohr radius of the hydrogen atom is $r = 1$.

The self-adjointness of such Hamiltonian operators, with domain $H^2(\mathbb{R}^{3N+3L})$, has been shown by Kato (1951); see also Reed & Simon (1975), Theorem X.16.

Any attempt to "solve" numerically the molecular Schrödinger equation

$$i\hbar \frac{\partial \Psi}{\partial t} = H_{\mathrm{mol}} \Psi , \qquad \Psi = \Psi(x_1, \dots, x_N, y_1, \dots, y_L, t)$$

faces a variety of severe problems:

- the high dimensionality (even for a small molecule such as CO_2, there are 3 nuclei and 22 electrons, so that Ψ is a function on $\mathbb{R}^{75}$!);
- multiple scales in the system (the mass of the electron is approximately 1/2000 of the mass of a proton);
- highly oscillatory wave functions.

To obtain satisfactory results in spite of these difficulties, one requires a combination of *model reduction*, based on physical insight and/or asymptotic analysis, and *numerical techniques* used on the reduced models that are intermediate between classical and full quantum dynamics. This is the subject of the following chapters.

Chapter II.
Reduced Models via Variational Approximation

There is a wide variety of models, or approximations, that are intermediate between the full time-dependent Schrödinger equation of many-body quantum mechanics and the Newtonian equations of classical mechanics. Most of these models are based on a time-dependent variational principle, first used by Dirac (1930), which plays a similarly fundamental role for the time-dependent Schrödinger equation as the Rayleigh-Ritz variational principle does for the Schrödinger eigenvalue problem. Indeed, several of the methods for the stationary problem, as for example the Hartree–Fock method, have a time-dependent analogue that comes about by the same choice of approximation manifold to which the variational principle is restricted. There are, however, different aspects that come into play in the time-dependent situation, both in the modeling/approximation aspects and in the numerical treatment of the reduced models.

We first give an abstract formulation and various interpretations of the time-dependent variational principle, and then turn to some basic examples that gradually take us from the full molecular Schrödinger equation down to classical molecular dynamics: the adiabatic or time-dependent Born–Oppenheimer approximation that eliminates the electronic degrees of freedom, the time-dependent self-consistent field approximation that separates the nuclei, and Gaussian wavepacket dynamics that parametrizes the single-particle wave functions. At the end of the chapter we address the theoretical question of quasi-optimality of variational approximations.

II.1 The Dirac–Frenkel Time-Dependent Variational Principle

In this section we give the abstract formulation of the time-dependent variational principle and discuss its structural properties.

II.1.1 Abstract Formulation

We consider an abstract Schrödinger equation on a complex Hilbert space $\mathcal{H}$ with inner product $\langle\cdot|\cdot\rangle$, taken antilinear in its first and linear in its second argument, and with a Hamiltonian H that is a self-adjoint linear operator on $\mathcal{H}$,

$$\frac{d\psi}{dt} = \frac{1}{i\hbar} H\psi\,. \tag{1.1}$$

Let $\mathcal{M}$ be a smooth submanifold of $\mathcal{H}$, and for $u \in \mathcal{M}$ denote by $\mathcal{T}_u\mathcal{M}$ the tangent space at u, which consists of the derivatives of all differentiable paths on $\mathcal{M}$ passing through u. We think of $\mathcal{M}$ as an *approximation manifold* on which an approximate solution $u(t)$ to the solution $\psi(t)$ of (1.1) with initial data $u(0) = \psi(0) \in \mathcal{M}$ is sought. The function $t \mapsto u(t) \in \mathcal{M}$ is determined from the condition that at every time t, its derivative $du/dt\,(t)$, which lies in the tangent space $\mathcal{T}_{u(t)}\mathcal{M}$, be such that the residual in the Schrödinger equation is orthogonal to the tangent space:

$$\frac{du}{dt} \in \mathcal{T}_u\mathcal{M} \quad \text{such that} \quad \Big\langle v \,\Big|\, \frac{du}{dt} - \frac{1}{i\hbar} Hu \Big\rangle = 0 \quad \forall\, v \in \mathcal{T}_u\mathcal{M}\,. \tag{1.2}$$

The tangent space $\mathcal{T}_u\mathcal{M}$ is known to be a real-linear closed subspace of $\mathcal{H}$. We will always assume that in fact

$$\mathcal{T}_u\mathcal{M} \;\text{ is a complex linear space,} \tag{1.3}$$

that is, with $v \in \mathcal{T}_u\mathcal{M}$, also $iv \in \mathcal{T}_u\mathcal{M}$. In this situation we get the same condition if we consider only the real part or the imaginary part of the inner product of (1.2). We will see, however, that these two cases lead to very different interpretations: as an orthogonal projection onto the tangent space in case of the real part, as a symplectic projection and as the Euler–Lagrange equations of an action functional in case of the imaginary part.

We remark that from a numerical analysis point of view, condition (1.2) can be seen as a Galerkin condition on the state-dependent approximation space $\mathcal{T}_u\mathcal{M}$.

Historical Note. Dirac (1930) used condition (1.2) without further comment to derive the equations of motion of what is now known as the time-dependent Hartree–Fock method. Frenkel (1934), p. 253, gives the interpretation as an orthogonal projection and refers to the appendix of the Russian translation of Dirac's book as the origin of the argument. Some thirty years later, the Dirac–Frenkel reasoning was taken up again by McLachlan (1964) and enriched by further examples. Condition (1.2) is therefore often called the Dirac–Frenkel–McLachlan time-dependent variational principle in the chemical physics literature, see Heller (1976) and, e.g., Baer & Billing (2002). In theoretical and nuclear physics, the derivation from Dirac's quantum-mechanical action functional and with it the symplectic viewpoint has rather been emphasized; see Kerman & Koonin (1976), Rowe, Ryman & Rosensteel (1980), Kramer & Saraceno (1981) and, e.g., Feldmeier & Schnack (2000).

II.1.2 Interpretation as an Orthogonal Projection

Taking the real part in (1.2), we arrive at the minimum condition for the following linear approximation problem:

$$\frac{du}{dt} \;\text{ is chosen as that } w \in \mathcal{T}_u\mathcal{M} \text{ for which } \Big\| w - \frac{1}{i\hbar} Hu \Big\| \;\text{ is minimal.} \tag{1.4}$$

(Note that $\|w + v - \frac{1}{i\hbar}Hu\|^2 = \|w - \frac{1}{i\hbar}Hu\|^2 + 2\,\mathrm{Re}\,\langle v \,|\, w - \frac{1}{i\hbar}Hu\rangle + \|v\|^2$.)

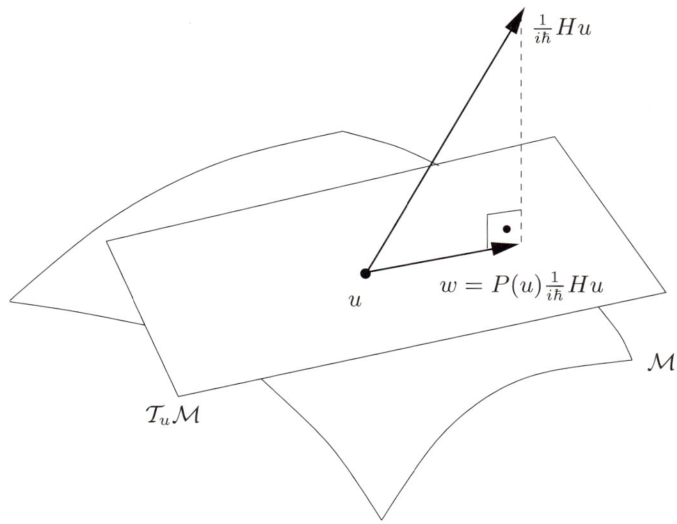

Fig. 1.1. Orthogonal projection to the tangent space.

In other words, du/dt is the *orthogonal projection* of $\frac{1}{i\hbar}Hu$ onto the tangent space $\mathcal{T}_u\mathcal{M}$. With the orthogonal projection operator onto $\mathcal{T}_u\mathcal{M}$ denoted by $P(u)$, we can thus rewrite (1.2) as a differential equation on the manifold $\mathcal{M}$,

$$\frac{du}{dt} = P(u)\frac{1}{i\hbar}Hu\,, \tag{1.5}$$

which is *nonlinear* unless $\mathcal{M}$ is a linear subspace of $\mathcal{H}$. The (global or local in time) existence of a solution $u(t) \in D(H) \cap \mathcal{M}$ can be ascertained only with further specifications about the operator H and the manifold $\mathcal{M}$. In the following we make formal calculations which implicitly assume that a sufficiently regular solution $u(t)$ exists.

II.1.3 Interpretation as a Symplectic Projection

The real-bilinear form

$$\omega(\xi,\eta) = -2\hbar\,\mathrm{Im}\,\langle \xi \,|\, \eta \rangle\,, \qquad \xi,\eta \in \mathcal{H}\,,$$

is antisymmetric, and ω is called the canonical *symplectic two-form* on $\mathcal{H}$. Since $\mathcal{T}_u\mathcal{M}$ is a complex linear space, for every $\varphi \in \mathcal{H}$ there exists a unique

$$w = P(u)\varphi \in \mathcal{T}_u\mathcal{M} \quad \text{such that} \quad \omega(v,w) = \omega(v,\varphi) \quad \forall v \in \mathcal{T}_u\mathcal{M}\,.$$

This non-degeneracy of the two-form ω makes $\mathcal{M}$ a *symplectic submanifold* of $\mathcal{H}$, and $P(u)$ is the *symplectic projection* operator onto $\mathcal{T}_u\mathcal{M}$. (Here $P(u)$ actually coincides with the orthogonal projection considered in the previous subsection.) Taking the imaginary part in condition (1.2) and multiplying with $-2\hbar$ yields

$$\omega\Big(v, \frac{du}{dt}\Big) = 2\,\mathrm{Re}\,\big\langle v \,\big|\, Hu \big\rangle \qquad \forall\, v \in \mathcal{T}_u\mathcal{M}\,. \tag{1.6}$$

We introduce the Hamilton function as the average of the Hamiltonian operator,

$$H(u) = \langle u \,|\, H \,|\, u \rangle$$

(we use the same symbol H as for the operator). The right-hand side in (1.6) is now recognized as the derivative $dH(u)v$ in the direction of v. Now, (1.6) rewritten as

$$\omega\Big(v, \frac{du}{dt}\Big) = dH(u)v \qquad \forall\, v \in \mathcal{T}_u\mathcal{M}\,, \tag{1.7}$$

is a *Hamiltonian system* on the symplectic manifold $\mathcal{M}$ with the Hamilton function $H(u)$; see Marsden & Ratiu (1999), Chap. 5.4. Let us state and verify basic properties of this system.

Theorem 1.1 (Energy Conservation). *The total energy $\langle H \rangle$ is conserved along solutions of the Hamiltonian system (1.7) on $\mathcal{M}$.*

Proof. We have (with $\dot{} = d/dt$)

$$\frac{d}{dt}\langle u \,|\, H \,|\, u \rangle = 2\,\mathrm{Re}\,\big\langle \dot u \,\big|\, Hu \big\rangle = \omega(\dot u, \dot u) = 0$$

on using (1.6) with $v = \dot u \in \mathcal{T}_u\mathcal{M}$ in the second equation. □

There is also the following important conservation property, which we first state briefly and then explain in detail.

Theorem 1.2 (Symplecticity). *The flow of the Hamiltonian system (1.7) is symplectic.*

This means that the symplectic two-form ω is preserved in the following sense: Let $u_0 \in \mathcal{M}$, and let $v_0 \in \mathcal{T}_{u_0}\mathcal{M}$ be a tangent vector at u_0. Then there is a path $\gamma(\tau)$ on $\mathcal{M}$ with $\gamma(0) = u_0$ and $d\gamma/d\tau\,(0) = v_0$. Let $u(t) = u(t, u_0)$ be the solution of (1.7) with initial data u_0, and denote by

$$v(t) = \frac{d}{d\tau}\Big|_{\tau=0} u(t, \gamma(\tau)) \in \mathcal{T}_{u(t)}\mathcal{M}$$

the tangent vector propagated along the solution $u(t, u_0)$ (note that $v(t)$ is the solution with initial data v_0 to the differential equation linearized at $u(t, u_0)$). Let $w(t)$ be another tangent vector propagated along the same solution, corresponding to an initial tangent vector w_0 at u_0. Then, the statement of Theorem 1.2 is that

$$\frac{d}{dt}\,\omega(v(t), w(t)) = 0\,. \tag{1.8}$$

Proof. By the bilinearity and antisymmetry of ω we have

$$\frac{d}{dt}\,\omega(v,w) = -\omega(w,\dot v) + \omega(v,\dot w)\,.$$

Differentiating (1.6) with respect to the inital value, we obtain that this equals

$$\frac{d}{dt}\,\omega(v,w) = -2\,\mathrm{Re}\,\big\langle w\,\big|\,Hv\big\rangle + 2\,\mathrm{Re}\,\big\langle v\,\big|\,Hw\big\rangle = 0\,. \qquad \square$$

We will further discuss symplectic and Hamiltonian aspects in Section II.4.2 where we consider the non-canonical Hamiltonian structure of the equations of motion for parametrized wave functions.

II.1.4 Interpretation as an Action Principle

Taking the imaginary part in (1.2) also yields that every solution of (1.2) makes the *action functional*

$$S(u) = \int_{t_0}^{t_1} \Big\langle u(t)\,\Big|\, i\hbar\frac{du}{dt}(t) - Hu(t)\Big\rangle\, dt \tag{1.9}$$

stationary with respect to variations of paths on the manifold $\mathcal{M}$ with fixed end-points, because by partial integration and the symmetry of H,

$$\begin{aligned}\delta S(u) &= \int_{t_0}^{t_1} \Big(\Big\langle \delta u(t)\,\Big|\, i\hbar\frac{du}{dt}(t) - Hu(t)\Big\rangle + \Big\langle u(t)\,\Big|\, i\hbar\frac{d\delta u}{dt}(t) - H\delta u(t)\Big\rangle\Big)\, dt \\ &= -2\hbar \int_{t_0}^{t_1} \mathrm{Im}\,\Big\langle \delta u(t)\,\Big|\, \frac{du}{dt}(t) - \frac{1}{i\hbar}Hu(t)\Big\rangle\, dt\,.\end{aligned}$$

The condition $\delta S = 0$ is the quantum-mechanical analogue of Hamilton's principle in classical mechanics. Also note that $S(u)$ is real if $\|u(t)\|^2 = \mathit{Const.}$, as is seen by partial integration in (1.9).

II.1.5 Conservation Properties

We know from the Heisenberg equation (I.4.4) that the average $\langle A\rangle$ is conserved along solutions of the Schrödinger equation if A commutes with the Hamiltonian H. For variational approximations (1.2) there is the following criterion.

Theorem 1.3 (Invariants). *Let the self-adjoint operator A commute with the Hamiltonian H, $[A,H]=0$. If*

$$Au \in \mathcal{T}_u\mathcal{M} \qquad \forall\, u \in \mathcal{M}\cap D(A)\,, \tag{1.10}$$

then the average of A along variational approximations $u(t) \in \mathcal{M}\cap D(A)$ is conserved: $\langle u(t)\,|\,A\,|\,u(t)\rangle = \mathit{Const.}$

Proof. We have

$$\frac{d}{dt}\langle u \,|\, A \,|\, u\rangle = 2\,\mathrm{Re}\,\langle Au \,|\, \dot{u}\rangle = 2\,\mathrm{Re}\,\langle Au \,|\, \frac{1}{i\hbar}\, Hu\rangle = \langle u \,|\, \frac{1}{i\hbar}\,[A,H] \,|\, u\rangle = 0$$

on using (1.2) and (1.10) in the second equality. □

Choosing A as the identity operator, we obtain the following useful corollary.

Theorem 1.4 (Norm Conservation). *The norm is conserved along variational approximations if $\mathcal{M}$ contains rays, that is, with $u \in \mathcal{M}$ also $\alpha u \in \mathcal{M}$ for all $\alpha > 0$.*

Proof. The stated condition implies $u \in \mathcal{T}_u\mathcal{M}$ for $u \in \mathcal{M}$, and hence the result follows from Theorem 1.3. □

II.1.6 An A Posteriori Error Bound

A simple but useful general error bound for variational approximations can be given in terms of the distance $\mathrm{dist}\,\big(\frac{1}{i\hbar}Hu, \mathcal{T}_u\mathcal{M}\big)$ in the Hilbert space norm of $\frac{1}{i\hbar}Hu$ along the variational approximation $u(t)$ to the corresponding tangent space. The error bound is *a posteriori* in that it is in terms of the approximation $u(t)$ rather than the exact wave function $\psi(t)$. This abstract result will be used in Sections II.4 (for Gaussian wave packets), III.1 (for spectral discretizations) and III.2 (for the Lanczos method).

Theorem 1.5 (Error Bound for Variational Approximations). *If $u(0) = \psi(0) \in \mathcal{M}$, then the error of the variational approximation is bounded by*

$$\|u(t) - \psi(t)\| \le \int_0^t \mathrm{dist}\left(\frac{1}{i\hbar}Hu(s), \mathcal{T}_{u(s)}\mathcal{M}\right) ds\,. \qquad (1.11)$$

Proof. We subtract (1.1) from (1.5), so that

$$\frac{d}{dt}(u-\psi) = \frac{1}{i\hbar}H(u-\psi) - P^\perp(u)\frac{1}{i\hbar}Hu \qquad \text{with} \quad P^\perp(u) = I - P(u)\,.$$

Multiplying with $u - \psi$ and taking the real part gives

$$\begin{aligned}\|u-\psi\| \cdot \frac{d}{dt}\,\|u-\psi\| &= \frac{1}{2}\,\frac{d}{dt}\,\|u-\psi\|^2 = \mathrm{Re}\,\big\langle u-\psi \,|\, \frac{d}{dt}(u-\psi)\big\rangle \\ &= \mathrm{Re}\,\big\langle u-\psi \,|\, -P^\perp(u)\frac{1}{i\hbar}Hu\big\rangle \le \|u-\psi\| \cdot \|P^\perp(u)\frac{1}{i\hbar}Hu\|\,.\end{aligned}$$

Dividing by $\|u-\psi\|$, integrating from 0 to t and noting

$$\mathrm{dist}\left(\frac{1}{i\hbar}Hu,\, \mathcal{T}_u\mathcal{M}\right) = \left\|P^\perp(u)\frac{1}{i\hbar}Hu\right\| = \left\|\frac{du}{dt} - \frac{1}{i\hbar}Hu\right\|$$

then yields the error bound (1.11). □

For the error in the average of an observable A along the variational approximation we note the bound

$$\big|\langle u \,|\, A \,|\, u\rangle - \langle \psi \,|\, A \,|\, \psi\rangle\big| = \big|\langle u-\psi \,|\, Au\rangle + \langle A\psi \,|\, u-\psi\rangle\big| \le \|u-\psi\| \cdot \big(\|Au\| + \|A\psi\|\big).$$

II.2 Adiabatic / Born–Oppenheimer Approximation

In the following three sections we turn to basic examples of variational approximation, which take us in steps from the full molecular Schrödinger equation down to classical molecular dynamics. We begin with the adiabatic approximation that separates the motion of heavy nuclei and light electrons.

II.2.1 Electronic Schrödinger Equation

We return to the molecular Hamiltonian (I.5.5), viz.,

$$H_{\mathrm{mol}} = T_N + T_e + V\,. \tag{2.1}$$

In a first step we ignore the contribution from the kinetic energy of the nuclei, T_N (vaguely motivated by the fact that $M_n \gg m$), and work with the electronic Hamiltonian

$$H_e(x) = T_e + V(x,\cdot)\,, \tag{2.2}$$

which acts on functions of the electronic coordinates $y = (y_1, \ldots, y_L)$ and depends parametrically on the nuclear coordinates $x = (x_1, \ldots, x_N)$. We consider the electronic structure problem, the Schrödinger eigenvalue problem

$$H_e(x)\Phi(x,\cdot) = E(x)\Phi(x,\cdot)\,, \tag{2.3}$$

typically for the smallest eigenvalue, the ground state energy. Actually computing eigenvalues and eigenfunctions of the electronic Schrödinger equation is the primary concern of computatonal quantum chemistry; see, e.g., Szabo & Ostlund (1996), and from a more mathematical viewpoint Le Bris (2003). Here we just suppose that this problem is solved in some satisfactory way.

We fix an eigenfunction $\Phi(x,\cdot)$ of $H_e(x)$ corresponding to the eigenvalue $E(x)$, and assume that $\Phi(x,y)$ is of unit L^2 norm as a function of y and depends smoothly on x. For fixed nuclear coordinates x, the solution of the *time-dependent electronic Schrödinger equation*

$$i\hbar\,\frac{\partial \Psi_e}{\partial t} = H_e(x)\Psi_e \tag{2.4}$$

with initial data $\psi_0(x)\Phi(x,\cdot)$ is given by

$$\Psi_e(x,y,t) = e^{-iE(x)t/\hbar}\psi_0(x)\cdot\Phi(x,y)\,. \tag{2.5}$$

II.2.2 Schrödinger Equation for the Nuclei on an Electronic Energy Surface

Equation (2.5) motivates the *adiabatic approximation* to the molecular Schrödinger equation, which is the variational approximation on

$$\mathcal{M} = \{u \in L^2_{x,y} : u(x,y) = \psi(x)\,\Phi(x,y),\ \psi \in L^2_x\}\,. \tag{2.6}$$

Here $L^2_x = L^2(\mathbb{R}^{3N})$ denotes the Lebesgue space of square integrable functions depending only on the nuclear coordinates x, and $L^2_{x,y} = L^2(\mathbb{R}^{3N} \times \mathbb{R}^{3L})$ is the L^2 space of functions depending on both nuclear and electronic coordinates. Note that here $\mathcal{M}$ is a complex linear space so that $\mathcal{T}_u\mathcal{M} = \mathcal{M}$ for all $u \in \mathcal{M}$. As we show below, the Dirac-Frenkel variational principle (1.2) then leads to a *Schrödinger equation for the nuclei on the electronic energy surface* E:

$$i\hbar\frac{\partial\psi}{\partial t} = H_N\psi \qquad \text{with} \quad H_N = T_N + E + B_1 + B_2\,, \tag{2.7}$$

$$B_1 = \sum_{n=1}^{N} \frac{\hbar}{M_n}\,\mathrm{Im}\,\langle \nabla_{x_n}\Phi \,|\, \Phi\rangle_{L^2_y} \cdot p_n\,, \qquad B_2 = \sum_{n=1}^{N} \frac{\hbar^2}{2M_n}\,\|\nabla_{x_n}\Phi\|^2_{L^2_y}\,,$$

with $p_n = -i\hbar\nabla_{x_n}$, where ∇_{x_n} is the gradient operator with respect to the variables x_n. The Hamiltonian H_N acts on functions of only the nuclear coordinates x, with the electronic eigenvalue E as a potential. The last two terms B_1 and B_2 contain derivatives of the electronic wave function Φ with respect to the nuclear coordinates x. They are usually neglected in computations, first because they are expensive to compute or simply not available and second by the formal argument – to be taken with caution – that they carry the large masses M_n in the denominator and are of lower differentiation order than the kinetic energy term. The resulting simplified approximation with the Hamiltonian

$$H_{\mathrm{BO}} = T_N + E \tag{2.8}$$

is known as the *time-dependent Born–Oppenheimer approximation*. It describes the motion of the nuclei as driven by the potential energy surface E of the electrons. It underlies the vast majority of computations in molecular dynamics.

The term B_2 can indeed be safely neglected since it can be shown that this omission introduces an error that is of the same magnitude as the approximation error in the adiabatic approximation.

The term B_1, known as the *Berry connection*, vanishes for real eigenfunctions Φ and, more generally, it can be made to vanish by a transformation $\Phi(x,y) \to e^{i\theta(x)}\Phi(x,y)$ with θ satisfying $\nabla_{x_n}\theta(x) = -\mathrm{Im}\,\langle\nabla_{x_n}\Phi \,|\, \Phi\rangle_{L^2_y}$. This transformation of Φ changes $\psi(x,t) \to e^{-i\theta(x)}\psi(x,t)$. Note that θ is uniquely determined up to a constant if Φ is indeed a smooth function of x on all of $\mathbb{R}^{3N}$, but is only locally uniquely determined if Φ is a differentiable function of x only on a domain that is not simply connected. In the latter case, B_1 can cause physical effects that are not retained in the model otherwise; see the extensive literature on Berry's phase, starting with Berry (1984) and Simon (1983).

Derivation of (2.7): We note that for $u(x,y) = \psi(x)\Phi(x,y)$ we have

$$T_N u = -\sum_{n=1}^{N} \frac{\hbar^2}{2M_n}\Big(\Delta_{x_n}\psi\cdot\Phi + 2\,\nabla_{x_n}\psi\cdot\nabla_{x_n}\Phi + \psi\cdot\Delta_{x_n}\Phi\Big),$$

and recall that $\|\Phi(x,\cdot)\|^2_{L^2_y} = 1$ for all x. We then obtain from (1.2) with $v(x,y) = \varphi(x)\Phi(x,y)$ for arbitrary $\varphi \in L^2_x$ that

$$\Big\langle \varphi \,\Big|\, i\hbar\frac{\partial\psi}{\partial t} - E\psi \;+\; \sum_{n=1}^{N} \frac{\hbar^2}{2M_n}\Big(\Delta_{x_n}\psi + 2\,\langle \nabla_{x_n}\Phi \,|\, \Phi\rangle_{L^2_y} \cdot \nabla_{x_n}\psi \\ - \langle \nabla_{x_n}\Phi \,|\, \nabla_{x_n}\Phi\rangle_{L^2_y}\,\psi\Big)\Big\rangle_{L^2_x} = 0\,.$$

On noting that $0 = \nabla_{x_n}\|\Phi\|^2_{L^2_y} = 2\,\mathrm{Re}\,\langle \nabla_{x_n}\Phi \,|\, \Phi\rangle_{L^2_y}$, we obtain (2.7). □

II.2.3 Semi-Classical Scaling

One property to the success of the adiabatic approximation is the smallness of the mass ratio of electrons and nuclei,

$$\varepsilon^2 = \frac{m}{M} \ll 1 \tag{2.9}$$

with $M = \min_n M_n$. For ease of presentation we assume in the following that the masses of the nuclei are all equal: $M_n = M$ for all n. In atomic units ($\hbar = 1$, $m = 1$, $r = 1$, $e = 1$) and with the small parameter ε of (2.9), the molecular Hamiltonian then takes the form

$$H^\varepsilon_{\mathrm{mol}} = -\frac{\varepsilon^2}{2}\Delta_x + H_e(x) \qquad \text{with} \quad H_e(x) = -\frac{1}{2}\Delta_y - V(x,\cdot)\,. \tag{2.10}$$

We are interested in solutions to the Schrödinger equation of bounded energy, and in particular of bounded kinetic energy

$$\langle \Psi \,|\, -\frac{\varepsilon^2}{2}\Delta_x \,|\, \Psi\rangle = \frac{1}{2}\,\|\varepsilon\nabla_x\Psi\|^2 = \mathcal{O}(1)\,.$$

For a wavepacket $e^{ip\cdot x}a(x)$ this condition corresponds to a momentum $p \sim \varepsilon^{-1}$ and hence to a velocity $v = p/M \sim \varepsilon$. Motion of the nuclei over a distance ~ 1 can thus be expected on a time scale ε^{-1}. We therefore rescale time

$$t \to t/\varepsilon\,,$$

so that with respect to the new time nuclear motion over distances ~ 1 can be expected to occur at time ~ 1. The molecular Schrödinger equation in the rescaled time then takes the form

$$i\varepsilon\frac{\partial\Psi}{\partial t} = H^\varepsilon_{\mathrm{mol}}\Psi\,. \tag{2.11}$$

The Schrödinger equation (2.7) for the nuclei becomes

$$i\varepsilon\frac{\partial\psi}{\partial t} = H^\varepsilon_N\psi \qquad \text{with} \quad H^\varepsilon_N = -\frac{\varepsilon^2}{2}\Delta_x + E + \varepsilon B_1 + \varepsilon^2 B_2\,, \tag{2.12}$$
$$B_1 = \mathrm{Im}\,\langle \nabla_x\Phi \,|\, \Phi\rangle_{L^2_y} \cdot p\,, \qquad B_2 = \frac{1}{2}\,\|\nabla_x\Phi\|^2_{L^2_y}\,,$$

with $p = -i\varepsilon\nabla_x$. We are interested in solutions over times $t = \mathcal{O}(1)$.

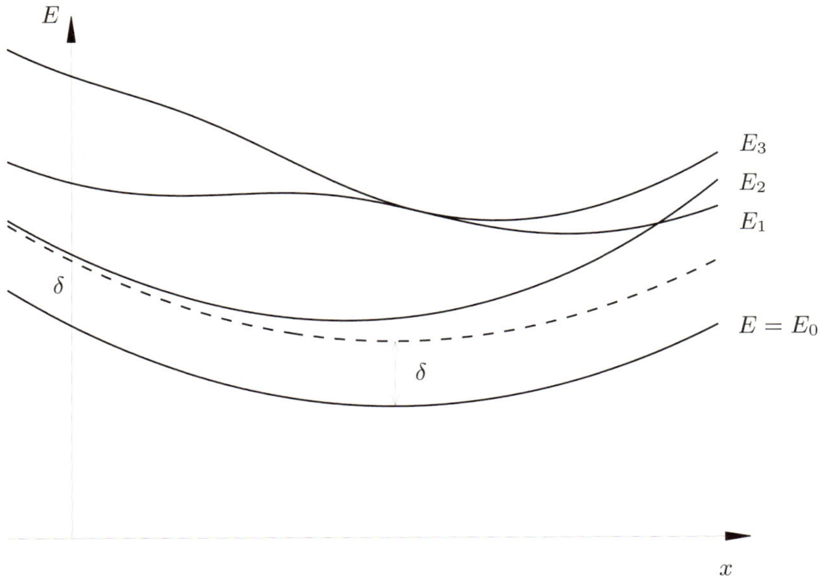

Fig. 2.1. Spectral gap.

II.2.4 Spectral Gap Condition

A small error of the adiabatic approximation will be seen to be caused by two properties: in addition to the smallness of the mass ratio $\varepsilon^2 = m/M$, we require a separation of the eigenvalue $E(x)$ from the remainder of the spectrum $\sigma(H_e(x))$ of the electronic Hamiltonian $H_e(x)$,

$$\operatorname{dist}\bigl(E(x),\, \sigma(H_e(x)) \setminus \{E(x)\}\bigr) \geq \delta > 0\,, \tag{2.13}$$

uniformly for all x in a region where the wavefunction remains approximately localized. We will give a result on the approximation error in the situation of a globally well-separated single eigenvalue $E(x)$, where (2.13) is assumed to hold uniformly for all $x \in \mathbb{R}^{3N}$.

Remark. It is known that the adiabatic approximation generally breaks down near crossings of eigenvalues. A remedy then is to enlarge the approximation space by including several energy bands that are well separated from the remaining ones in the region of physical interest, e.g., using

$$\mathcal{M} = \{u \in L^2_{x,y} : u(x,y) = \psi_1(x)\Phi_1(x,y) + \psi_2(x)\Phi_2(x,y),\ \psi_1, \psi_2 \in L^2_x\}\,, \tag{2.14}$$

where $\Phi_1(x,\cdot)$, $\Phi_2(x,\cdot)$ span an invariant subspace of the electronic Hamiltonian $H_e(x)$. The variational approximation on $\mathcal{M}$ then leads to a system of coupled Schrödinger equations:

$$i\hbar \frac{\partial \psi}{\partial t} = T_N \psi + B_1 \psi + B_2 \psi + V \psi \qquad \text{for} \quad \psi = \begin{pmatrix} \psi_1 \\ \psi_2 \end{pmatrix} \tag{2.15}$$

with the matrix-valued potential

$$V = \begin{pmatrix} V_{11} & V_{12} \\ V_{21} & V_{22} \end{pmatrix} \qquad \text{with} \quad V_{ij}(x) = \langle \Phi_i(x,\cdot) \,|\, H_e(x) \,|\, \Phi_j(x,\cdot) \rangle_{L^2_y} \tag{2.16}$$

and with the diagonal operators $B_j = \begin{pmatrix} B_j^1 & 0 \\ 0 & B_j^2 \end{pmatrix}$, where B_j^k are defined as B_j in (2.7) with Φ_k instead of Φ.

The non-adiabatic solution behaviour near eigenvalue crossings has attracted much attention in recent years; see, e.g., Baer & Billing (2002), Domcke, Yarkony & Köppel (2004), and Lasser & Teufel (2005).

II.2.5 Approximation Error

We derive an error bound of the adiabatic approximation that works for a modified Hamiltonian where the Coulomb interactions of the nuclei are mollified to smooth bounded potentials. We assume

$$\|\nabla_x V(x,y)\| \le C_V \qquad \text{for} \quad x \in \mathbb{R}^{3N},\ y \in \mathbb{R}^{3L} \tag{2.17}$$

and consider initial data on the approximation space $\mathcal{M}$ of (2.6),

$$\Psi_0(x,y) = \psi_0(x)\Phi(x,y) \qquad \text{with} \quad \|H_N^\varepsilon \psi_0\| \le C_0\,,\ \|\psi_0\| = 1\,. \tag{2.18}$$

We consider the adiabatic approximation $u(t) = u(\cdot,\cdot,t)$, with initial data Ψ_0, determined by the time-dependent variational principle:

$$\frac{\partial u}{\partial t} \in \mathcal{M} \qquad \text{such that} \quad \Big\langle v \,\Big|\, \frac{\partial u}{\partial t} - \frac{1}{i\varepsilon} H^\varepsilon_{\mathrm{mol}} u \Big\rangle = 0 \quad \forall\, v \in \mathcal{M}\,. \tag{2.19}$$

We know already that

$$u(x,y,t) = \psi(x,t)\Phi(x,y)\,, \tag{2.20}$$

where $\psi(x,t)$ is the solution of the nuclear Schrödinger equation (2.12) with initial data $\psi_0(x)$. This is compared with the exact solution $\Psi(t) = \Psi(\cdot,\cdot,t)$ of the molecular Schrödinger equation (2.11) with initial data $\Psi_0(x,y) = \psi_0(x)\Phi(x,y)$.

Theorem 2.1 (Space-Adiabatic Theorem, Teufel 2003). *Under the above conditions, the error of the adiabatic approximation is bounded by*

$$\|u(t) - \Psi(t)\| \le C\,(1+t)\,\varepsilon \qquad \textit{for} \quad t \ge 0\,,$$

where C is independent of ε and t and initial data satisfying (2.18), but depends on the gap δ of (2.13) (uniform for $x \in \mathbb{R}^{3N}$), on bounds of partial derivatives with respect to x up to third order of the eigenfunctions Φ, and on the bounds C_V of (2.17) and C_0 of (2.18).

Teufel (2003) gives a more general result, including the case of higher-dimensional invariant subspaces as in (2.14), and a wealth of related theory. In the global version stated above, the result remains valid for the time-dependent Born–Oppenheimer approximation (2.8), with the same proof after eliminating B_1 by the gauge transformation discussed after (2.8). This is no longer true for local versions of the result where the eigenfunction Φ is defined only on a domain that is not simply connected.

The result is also related to the time-adiabatic theorem of Born & Fock (1928) and Kato (1950), which states that in a quantum system with a slowly time-varying Hamiltonian a wave function that is an eigenfunction initially, approximately remains an eigenfunction of the Hamiltonian at any instant for long times.

Proof. We let $H = H^{\varepsilon}_{\mathrm{mol}}$ for brevity, so that $\Psi(t)$ is a solution to

$$i\varepsilon \frac{\partial \Psi}{\partial t} = H\Psi .$$

With the orthogonal projection P onto $\mathcal{M}$, we reformulate (2.19) as

$$i\varepsilon \frac{\partial u}{\partial t} = Ku \qquad \text{with} \quad K = PHP^* ,$$

noting that K is a self-adjoint operator on $\mathcal{M}$. We then have $u(t) = e^{-itK/\varepsilon}\Psi_0 = Pe^{-itK/\varepsilon}\Psi_0 \in \mathcal{M}$, and by the variation-of-constants formula for $i\varepsilon\frac{\partial}{\partial t}(u - \Psi) = H(u-\Psi) - (H-K)u$,

$$\begin{aligned} u(t) - \Psi(t) &= -\frac{1}{i\varepsilon}\int_0^t e^{-i(t-s)H/\varepsilon}(H-K)u(s)\,ds \\ &= -\frac{1}{i\varepsilon}\int_0^t e^{-i(t-s)H/\varepsilon}(H-K)Pe^{-isK/\varepsilon}\Psi_0\,ds . \end{aligned}$$

We note that $(H-K)P = P^{\perp}HP$ (with $P^{\perp} = I - P$ the complementary orthogonal projection). The key idea is now to write $P^{\perp}HP$ essentially as a commutator with H, which becomes possible by the gap condition (2.13). Lemma 2.2 below tells us that $P^{\perp}HP = \varepsilon[H,G] + \varepsilon^2 R$ with operators G and R that are bounded independently of ε in appropriate norms as stated there. The remainder term $\varepsilon^2 R$ immediately gives an $\mathcal{O}(\varepsilon)$ bound on time intervals of length $\mathcal{O}(1)$ as desired. We then have

$$u(t) - \Psi(t) = ie^{-itH/\varepsilon}\int_0^t e^{isH/\varepsilon}\,[H,G]\,e^{-isH/\varepsilon}\cdot e^{isH/\varepsilon}\,e^{-isK/\varepsilon}\,\Psi_0\,ds + \mathcal{O}(t\varepsilon) ,$$

where we observe the key relation to gain a factor ε,

$$e^{isH/\varepsilon}\,[H,G]e^{-isH/\varepsilon} = -i\varepsilon\,\frac{d}{ds}\Big(e^{isH/\varepsilon}\,G\,e^{-isH/\varepsilon}\Big) .$$

We now use partial integration and note

$$\frac{d}{ds}\Big(e^{isH/\varepsilon}\,e^{-isK/\varepsilon}\Psi_0\Big) = \frac{i}{\varepsilon}e^{isH/\varepsilon}\,(H-K)P\,e^{-isK/\varepsilon}\Psi_0 .$$

Expressing once again $(H - K)P = P^\perp HP$, we obtain

$$u(t) - \Psi(t) = \varepsilon\, Ge^{-itK/\varepsilon}\,\Psi_0 - \varepsilon\, e^{-itH/\varepsilon} G\Psi_0 \\ - i\int_0^t e^{-i(t-s)H/\varepsilon}\, GP^\perp HP\, e^{-isK/\varepsilon}\Psi_0\, ds + \mathcal{O}(t\varepsilon)\,. \tag{2.21}$$

The result now follows with the estimates of Lemmas 2.2 and 2.3. □

It remains to state and prove the two lemmas to which we referred in the above proof. They use scaled Sobolev norms of functions on $\mathbb{R}^{3N}$ or $\mathbb{R}^{3N} \times \mathbb{R}^{3L}$. The squares of these norms are defined by

$$\|\varphi\|_{1,\varepsilon}^2 = \|\varepsilon\nabla_x\varphi\|^2 + \|\varphi\|^2\,,$$
$$\|\varphi\|_{2,\varepsilon}^2 = \|\varepsilon^2\Delta_x\varphi\|^2 + \|\varphi\|^2\,,$$

where the norm on the right-hand side is the L^2 norm (the L^2_x or $L^2_{x,y}$ norm, as appropriate).

Lemma 2.2. *The projected Hamiltonian $P^\perp HP$ can be written as*

$$P^\perp HP = \varepsilon[H, G] + \varepsilon^2 R \tag{2.22}$$

where the operators G and R are bounded by

$$\|G\Psi\| \le C_1\, \|\Psi\|_{1,\varepsilon}\,, \qquad \|R\Psi\| \le C_2\, \|\Psi\|_{2,\varepsilon} \tag{2.23}$$

for all $\Psi \in C_0^\infty(\mathbb{R}^{3N} \times \mathbb{R}^{3L})$. Moreover, $P^\perp HP$ is bounded by

$$\|P^\perp HP\Psi\|_{1,\varepsilon} \le C\varepsilon\, \|\Psi\|_{2,\varepsilon}\,. \tag{2.24}$$

Proof. In the following we write $\nabla = \nabla_x$ and $\Delta = \Delta_x$ for the gradient and Laplacian with respect to the nuclear coordinates x.

(a) We begin by computing $P^\perp HP$ for $H = -\frac{\varepsilon^2}{2}\Delta + H_e$. The orthogonal projection P onto $\mathcal{M}$ is fibered as

$$(P\Psi)(x) = P(x)\Psi(x,\cdot)\,,$$

where $P(x)$ is the L^2_y-orthogonal projection onto the span of the eigenfunction $\Phi(x,\cdot)$ of the electronic Hamiltonian $H_e(x)$. We have, for $\eta \in L^2_y$,

$$P(x)\eta = \langle \Phi(x,\cdot)\,|\,\eta\rangle\, \Phi(x,\cdot)\,,$$

with the inner product of L^2_y. Since $\Phi(x,\cdot)$ spans an invariant subspace of $H_e(x)$, we have $P^\perp(x)H_e(x)P(x) = 0$, and hence, for $\Psi \in L^2_{x,y}$,

$$P^\perp HP\Psi = -\frac{\varepsilon^2}{2}\, P^\perp\Delta(P\Psi) = -\varepsilon^2 P^\perp(\nabla P)\cdot\nabla\Psi - \frac{\varepsilon^2}{2}\, P^\perp(\Delta P)\Psi\,.$$

For the first term on the right-hand side we note, using $(\nabla P)P^{\perp}\Psi = \langle \nabla\Phi \,|\, P^{\perp}\Psi\rangle \Phi$ and $P^{\perp}\Phi = 0$,

$$Q := -P^{\perp}(\nabla P) = -P^{\perp}(\nabla P)P\,.$$

We thus obtain

$$P^{\perp}HP = \varepsilon\, Q\cdot\varepsilon\nabla + \varepsilon^2 R_0\,, \tag{2.25}$$

where $R_0(x) = -\frac{1}{2}P(x)^{\perp}(\Delta P)(x)$ is bounded on L^2_y uniformly in $x\in\mathbb{R}^{3N}$, provided that the eigenfunction Φ has bounded derivatives with respect to x. We also note that (2.25) implies the bound (2.24).

(b) We construct $F(x)$ such that

$$[H_e(x), F(x)] = Q(x)\,. \tag{2.26}$$

Writing H_e as an operator matrix with blocks corresponding to $\mathcal{M}$ and $\mathcal{M}^{\perp}$,

$$H_e = \begin{pmatrix} E & 0 \\ 0 & H_e^{\perp} \end{pmatrix} \quad \text{with} \quad H_e^{\perp} = P^{\perp}H_eP^{\perp}\,,$$

we can rewrite (2.26) as

$$\left[\begin{pmatrix} E & 0 \\ 0 & H_e^{\perp} \end{pmatrix}, \begin{pmatrix} F_{11} & F_{12} \\ F_{21} & F_{22} \end{pmatrix}\right] = \begin{pmatrix} 0 & 0 \\ Q_{21} & 0 \end{pmatrix}$$

which is solved by setting $F_{11}=0$, $F_{12}=0$, $F_{22}=0$ and determining $F_{21} = P^{\perp}FP$ from

$$H_e^{\perp}F_{21} - F_{21}E = Q_{21}\,.$$

By the spectral gap condition (2.13), this equation has a unique solution, and we thus obtain the solution to (2.26) as

$$F(x) = \bigl(H_e^{\perp}(x) - E(x)\bigr)^{-1} Q(x)\,.$$

This is bounded in L^2_y uniformly for $x\in\mathbb{R}^{3N}$ by the uniform gap condition, and so are $\nabla F(x)$ and $\Delta F(x)$.

(c) We next show that the commutator of $H = -\frac{\varepsilon^2}{2}\Delta + H_e$ with F is a small perturbation to $[H_e, F] = Q$. For this we note that

$$\Bigl[-\frac{\varepsilon^2}{2}\Delta, F\Bigr] = -\varepsilon\nabla F\cdot\varepsilon\nabla - \frac{\varepsilon^2}{2}\,\Delta F(x)\,,$$

so that

$$[H, F] = Q - \varepsilon R_1\,, \tag{2.27}$$

where R_1 is bounded by $\|R_1\Psi\| \le c_1\|\Psi\|_{1,\varepsilon}$ for all Ψ.

(d) We set

$$G = F\cdot\varepsilon\nabla \tag{2.28}$$

and show that the commutator with H equals $Q \cdot \varepsilon\nabla$ up to a small perturbation. By (2.27) we have, using the Leibniz rule of the commutator,

$$\begin{aligned}[H,G] &= [H,F]\cdot\varepsilon\nabla + F\cdot[H,\varepsilon\nabla] \\ &= Q\cdot\varepsilon\nabla - \varepsilon R_1\cdot\varepsilon\nabla - \varepsilon F\cdot\nabla V\,.\end{aligned}$$

For the term with the potential V we recall assumption (2.17), which bounds ∇V. The term $Q\cdot\varepsilon\nabla$ is the same as in (2.25), and hence we obtain the desired result (2.22) with $R = R_0 + R_1\cdot\varepsilon\nabla + F\cdot\nabla V$. The bounds (2.23) are immediate from the construction of the operators G and R. □

We also need the following regularity result in order to use (2.24) in (2.21).

Lemma 2.3. *In the situation of Theorem 2.1, we have*

$$\|u(t)\|_{2,\varepsilon} \le C\left(\|H_N^\varepsilon\psi_0\| + 1\right) \qquad \textit{for} \quad t \ge 0\,.$$

Proof. We use the bounds, for $\psi\Phi \in \mathcal{M}$,

$$\|\psi\Phi\|_{2,\varepsilon} \le c\,\|\psi\|_{2,\varepsilon} \le C\left(\|H_N^\varepsilon\psi\| + \|\psi\|\right),$$

for which we omit the straightforward derivation. By (2.20), the adiabatic approximation is given as $u(t) = \left(e^{-itH_N^\varepsilon/\varepsilon}\psi_0\right)\Phi$, and the above inequality thus yields

$$\|u(t)\| \le C\big(\|H_N^\varepsilon e^{-itH_N^\varepsilon/\varepsilon}\psi_0\| + \|\psi_0\|\big) = C\big(\|H_N^\varepsilon\psi_0\| + 1\big)\,,$$

which is the stated bound. □

II.3 Separating the Particles: Self-Consistent Field Methods

The remaining high dimensionality requires further model reductions. The many-body wave function is approximated by appropriate linear combinations of tensor products of single-particle wave functions. The simplest case arises in approximating the dynamics of the nuclei by a single tensor product, which yields the *time-dependent Hartree method*. This model describes the motion of each particle driven by the mean field of the other particles.

Its antisymmetrized version, suitable for electron dynamics, is known as the *time-dependent Hartree–Fock method*. The equations of motion for the orbitals were derived by Dirac (1930) in what is the historically first application of the time-dependent variational principle. This method is the time-dependent counterpart of the stationary Hartree–Fock method, which uses antisymmetrized products of orbitals to approximate eigenfunctions of the Schrödinger operator and is the basic approach to electronic structure computations; see, e.g., Szabo & Ostlund (1996).

Taking linear combinations of tensor products or their antisymmetrizations yields the *multi-configuration* time-dependent Hartree and Hartree–Fock methods, put forward by Meyer, Manthe & Cederbaum (1990). In this section we describe these various methods, derive the nonlinear equations of motion and discuss some of their properties.

The model reductions of this section can be viewed as *low-rank approximations* to the high-dimensional multi-particle wave function. Independently of the developments in quantum mechanics, low-rank approximations to huge matrices and tensors have been widely used as computationally viable approximations in many other fields including, for example, information retrieval, image compression, and option pricing. It seems, however, that using the time-dependent variational principle for low-rank approximations in areas outside quantum mechanics has been considered only recently (Koch & Lubich 2007, Nonnenmacher & Lubich 2008, Jahnke & Huisinga 2007).

II.3.1 Time-Dependent Hartree Method (TDH)

We consider the Schrödinger equation for the nuclei obtained from the Born–Oppenheiner approximation,

$$i\hbar\frac{\partial\psi}{\partial t} = H\psi\,, \qquad H = T + V \tag{3.1}$$

with kinetic energy $T = -\sum_{n=1}^{N} \frac{\hbar^2}{2M_n}\,\Delta_{x_n}$ and a potential $V(x_1,\ldots,x_N)$ (as an approximation to an electronic energy surface $E(x_1,\ldots,x_N)$). We assume that the domain $D(V)$ contains $D(T) = H^2(\mathbb{R}^{3N})$.

Hartree Products. We look for an approximation to the wave function of the tensor product form

$$\psi(x_1,\ldots,x_N,t) \approx a(t)\,\varphi_1(x_1,t)\ldots\varphi_N(x_N,t)$$

with a scalar phase factor $a(t)$ and with *single-particle functions* (or *molecular orbitals*) $\varphi_n(x_n,t)$. We thus consider the variational approximation (1.2) on the infinite-dimensional manifold

$$\mathcal{M} = \{u \in L^2(\mathbb{R}^{3N}) \,:\, u \neq 0,\ u = a\,\varphi_1 \otimes \ldots \otimes \varphi_N\,,\ a \in \mathbb{C},\ \varphi_n \in L^2(\mathbb{R}^3)\} \tag{3.2}$$

(or instead we might consider tensor products of $3N$ functions in $L^2(\mathbb{R})$). The representation of $u \in \mathcal{M}$ as $u = a\,\varphi_1 \otimes \ldots \otimes \varphi_N$ is not unique: for any choice of complex numbers $c_n \neq 0$, u remains unaltered under the transformation

$$\varphi_n \to c_n\varphi_n\,, \qquad a \to \frac{a}{c_1\ldots c_N}\,. \tag{3.3}$$

Tangent Functions. Although we do not have a unique representation of functions in the Hartree manifold $\mathcal{M}$, we can obtain a unique representation of tangent functions. This is what matters in deriving the equations of motion for the single-particle functions. Consider $u = a\,\varphi_1 \otimes \ldots \otimes \varphi_N$ with a of unit modulus and all φ_n of unit L^2 norm. Every tangent function $\dot{u} \in \mathcal{T}_u\mathcal{M}$ (for the moment, $\dot{u}$ is just a symbol for any tangent function) is of the form

$$\dot{u} = \dot{a}\,\varphi_1 \otimes \ldots \otimes \varphi_N + a\,\dot{\varphi}_1 \otimes \varphi_2 \otimes \ldots \otimes \varphi_N + \ldots + a\,\varphi_1 \otimes \ldots \otimes \varphi_{N-1} \otimes \dot{\varphi}_N \quad (3.4)$$

where $\dot{a} \in \mathbb{C}$ and $\dot{\varphi}_n \in L^2$. These turn out to be uniquely determined by $\dot{u}$ and the fixed $a, \varphi_1, \ldots, \varphi_N$ if we impose the *gauge condition*

$$\langle \varphi_n \,|\, \dot{\varphi}_n \rangle = 0\,. \quad (3.5)$$

Indeed, taking the inner product of both sides of (3.4) with $u = a\,\varphi_1 \otimes \ldots \otimes \varphi_N$ and using (3.5) and $\|\varphi_n\| = 1$ and $a = 1/\overline{a}$, determines $\dot{a}$ as

$$\dot{a} = \langle u \,|\, \dot{u} \rangle\, a\,. \quad (3.6)$$

Taking the inner product with the function in which the nth factor φ_n in u is replaced by some L^2 function ϑ_n, viz., with $a\,\varphi_1 \otimes \ldots \otimes \vartheta_n \otimes \ldots \otimes \varphi_N \in \mathcal{T}_u\mathcal{M}$, determines $\dot{\varphi}_n$ uniquely by the equation

$$\langle \vartheta_n \,|\, \dot{\varphi}_n \rangle + \overline{a}\dot{a}\langle \vartheta_n \,|\, \varphi_n \rangle = \langle a\,\varphi_1 \otimes \ldots \otimes \vartheta_n \otimes \ldots \otimes \varphi_N \,|\, \dot{u} \rangle \qquad \forall\, \vartheta_n \in L^2\,. \quad (3.7)$$

Equations of Motion for the Single-Particle Functions. We now consider the variational approximation (1.2) on the Hartree manifold $\mathcal{M}$, viz.,

$$\Big\langle v \,\Big|\, \frac{du}{dt} - \frac{1}{i\hbar} H u \Big\rangle = 0 \qquad \forall\, v \in \mathcal{T}_u\mathcal{M}\,. \quad (3.8)$$

Applying the above argument with $\dot{u} = du/dt \in \mathcal{T}_u\mathcal{M}$ and using (3.8) to replace $\dot{u}$ by $\frac{1}{i\hbar}Hu$ in (3.6) and (3.7), we obtain evolution equations for the factors in $u = a\,\varphi_1 \otimes \ldots \otimes \varphi_N$:

$$\begin{aligned} \frac{da}{dt} &= \Big\langle u \,\Big|\, \frac{1}{i\hbar} H u \Big\rangle\, a \\ \Big\langle \vartheta_n \,\Big|\, \frac{\partial \varphi_n}{\partial t} \Big\rangle &= \Big\langle a\,\varphi_1 \otimes \ldots \otimes \vartheta_n \otimes \ldots \otimes \varphi_N \,\Big|\, \frac{1}{i\hbar} H u \Big\rangle \\ &\quad - \Big\langle u \,\Big|\, \frac{1}{i\hbar} H u \Big\rangle \langle \vartheta_n \,|\, \varphi_n \rangle \qquad\qquad \forall\, \vartheta_n \in L^2\,. \end{aligned} \quad (3.9)$$

With the total energy $\kappa = \langle u \,|\, H \,|\, u \rangle$, which by Theorem 1.1 is constant in time, and with the *mean-field Hamiltonian* for the nth particle,

$$\langle H \rangle_n = \langle \psi_n \,|\, H \,|\, \psi_n \rangle \qquad \text{with} \quad \psi_n = \bigotimes_{j \neq n} \varphi_j \quad (3.10)$$

(the inner product on the right-hand side is over all variables except x_n), the equations of motion become the trivial linear constant-coefficient differential equation $i\hbar\, da/dt = \kappa a$ and

$$i\hbar \frac{\partial \varphi_n}{\partial t} = \langle H \rangle_n \varphi_n - \kappa \varphi_n \, . \tag{3.11}$$

Multiplying with $\frac{1}{i\hbar}\varphi_n$ and noting

$$\frac{d}{dt} \|\varphi_n\|^2 = 2\,\mathrm{Re} \left\langle \varphi_n \,\middle|\, \frac{\partial \varphi_n}{\partial t} \right\rangle = 2\,\mathrm{Re} \left\langle \varphi_n \,\middle|\, \frac{1}{i\hbar}\big(\langle H \rangle_n \varphi_n - \kappa \varphi_n\big) \right\rangle = 0 \, ,$$

we see that φ_n indeed remains of unit norm, as was assumed in the derivation.

The last term $\kappa\varphi_n$ in (3.11) can be dropped if we rescale $\varphi_j \to e^{-i\kappa t/\hbar}\varphi_j$. For a Hamiltonian $H = T + V$ as in (3.1), we obtain for all $\vartheta_n \in L^2(\mathbb{R}^3)$ that are orthogonal to φ_n,

$$\left\langle \varphi_1 \otimes \ldots \otimes \vartheta_n \otimes \ldots \otimes \varphi_N \,\middle|\, Tu \right\rangle = \left\langle \vartheta_n \,\middle|\, -\frac{\hbar^2}{2M_n} \Delta_{x_n} \varphi_n \right\rangle ,$$

and hence for such ϑ_n we have by (3.9)

$$\left\langle \vartheta_n \,\middle|\, i\hbar \frac{\partial \varphi_n}{\partial t} + \frac{\hbar^2}{2M_n} \Delta_{x_n} \varphi_n - \langle V \rangle_n \varphi_n \right\rangle = 0 \, ,$$

where the mean-field potential $\langle V \rangle_n$ is defined in the same way as in (3.10) with V instead of H. It follows that the right-hand expression in the inner product is a multiple of φ_n. This term adds to $\dot{u} = du/dt$ in (3.4) only a scalar multiple of u and hence yields only a modified phase factor a in u. Let us summarize the result obtained.

Theorem 3.1 (Time-Dependent Hartree Method). *For a Hamiltonian (3.1), the variational approximation on the Hartree manifold (3.2), for initial data* $u(x_1, \ldots, x_N, 0) = \varphi_1(x_1, 0) \ldots \varphi_N(x_N, 0)$ *with* $\varphi_n(\cdot, 0)$ *of unit* L^2 *norm, is given as*

$$u(x_1, \ldots, x_N, t) = a(t)\, \varphi_1(x_1, t) \ldots \varphi_N(x_N, t) \, ,$$

where $|a(t)| = 1$ *and* $\varphi_n(x_n, t)$ *are solutions to the system of nonlinear partial differential equations*

$$i\hbar \frac{\partial \varphi_n}{\partial t} = -\frac{\hbar^2}{2M_n} \Delta_{x_n} \varphi_n + \langle V \rangle_n \varphi_n \, . \tag{3.12}$$

This holds on time intervals $0 \le t \le \bar{t}$ *on which a strong solution to this system exists, that is, for* $\varphi_n \in C^1([0, \bar{t}], L^2(\mathbb{R}^3)) \cap C([0, \bar{t}], H^2(\mathbb{R}^3))$. □

The mean-field potentials $\langle V \rangle_n$ are high-dimensional integrals. Their computation is reduced to low-dimensional integrals for potentials that are (or more often, are approximated by) a linear combination of tensor products,

$$V(x_1, \ldots, x_N) = \sum_{k=1}^{r} c_k \, v_k^{(1)}(x_1) \cdot \ldots \cdot v_k^{(N)}(x_N) \, , \tag{3.13}$$

for which

$$\langle V \rangle_n = \sum_{k=1}^{r} c_k \prod_{j \ne n} \int_{\mathbb{R}^3} v_k^{(j)}(x_j) \, |\varphi_j(x_j)|^2 \, dx_j \, .$$

Equations (3.12) look like usual Schrödinger equations, but since the mean-field potential $\langle V \rangle_n$ depends on the single-particle functions of the other particles, we obtain a coupled system of low-dimensional *nonlinear* partial differential equations. A strong solution to (3.12) exists globally for all times $t \geq 0$ for example in the case of a smooth bounded potential with bounded derivatives. This is shown by Picard iteration in the Sobolev space $H^2(\mathbb{R}^3)^N$ on the integrated equations

$$\varphi_n(t) = e^{-itT_n/\hbar}\varphi_n(0) + \int_0^t e^{-i(t-s)T_n/\hbar} \langle V \rangle_n(s)\, \varphi_n(s)\, ds \,,$$

where $T_n = -\frac{\hbar^2}{2M_n}\Delta_{x_n}$. By the same argument, the solution then has H^k regularity for arbitrary k whenever the initial data is in H^k.

Remark 3.2 (Principal Bundle Structure). On the Hartree manifold $\mathcal{M}$ of (3.2), $y = (a, \varphi_1, \ldots, \varphi_N)$ are not coordinates, but the underlying mathematical structure here and in the following subsections is that of a *principal bundle*, which is a familar concept in differential geometry that we now describe. There is a map $\chi : \mathcal{N} \to \mathcal{M}$ from a manifold $\mathcal{N}$ onto $\mathcal{M}$, so that every $u \in \mathcal{M}$ can be represented, though not uniquely, as

$$u = \chi(y) \qquad \text{for some } y \in \mathcal{N}.$$

(We have $\chi(y) = a\,\varphi_1 \otimes \ldots \otimes \varphi_N$ on the Hartree manifold.) The map χ is invariant under the action of a Lie group G on $\mathcal{N}$, which we denote by $\cdot : G \times \mathcal{N} \to \mathcal{N}$:

$$\chi(g \cdot y) = \chi(y) \qquad \forall g \in G\,,\; y \in \mathcal{N}\,.$$

In the Hartree method, the group is the componentwise multiplicative group $G = (\mathbb{C}^*)^N$ (with $\mathbb{C}^* = \mathbb{C} \setminus \{0\}$), and the action is given by (3.3).

Moreover, there is a *gauge map* γ, which at every $y \in \mathcal{N}$ associates to a tangent vector $\dot{y} \in T_y\mathcal{N}$ an element $\gamma(y)\dot{y}$ in the Lie algebra $\mathfrak{g}$ of G ($\mathfrak{g}$ is the tangent space at the unit element of G). The linear map $\gamma(y) : T_y\mathcal{N} \to \mathfrak{g}$ is such that the extended derivative map, with $u = \chi(y)$,

$$T_y\mathcal{N} \to T_u\mathcal{M} \times \mathfrak{g} \,:\, \dot{y} \mapsto \big(d\chi(y)\dot{y},\, \gamma(y)\dot{y}\big) \qquad \text{is an isomorphism.}$$

Hence, under the gauge condition $\gamma(y)\dot{y} = 0$ (or with any fixed element of $\mathfrak{g}$ instead of 0), $\dot{y} \in T_y\mathcal{N}$ is determined uniquely by y and $\dot{u} \in T_u\mathcal{M}$. In the Hartree method, a gauge map is given by $\gamma(y)\dot{y} = \big(\langle \varphi_n \,|\, \dot{\varphi}_n \rangle\big)_{n=1}^N \in \mathbb{C}^N$.

II.3.2 Time-Dependent Hartree–Fock Method (TDHF)

Slater Determinants. For a system of N identical fermions[1] the wave function is antisymmetric (see Sect. I.5.2) and we wish to retain this property in the approximation. We

[1] Typically, this refers to electrons. In the notation of Sect. I.5.3, their coordinates are $y_1, \ldots, y_L$, but we will denote them by $x_1, \ldots, x_N$ in this subsection and keep in mind that the x_n are not coordinates of nuclei here.

therefore look for an approximate wave function in an antisymmetrized tensor product form, that is, as a *Slater determinant*

$$\psi(x_1,\dots,x_N,t) \approx a(t)\frac{1}{\sqrt{N!}}\det\big(\varphi_n(x_j,t)\big)_{n,j=1}^{N}$$

with a scalar phase factor $a(t)$ and with orbitals $\varphi_n(x,t)$ that are time-dependent functions of $x \in \mathbb{R}^3$. In the following we write the scaled determinant as the wedge product

$$\varphi_1 \wedge \ldots \wedge \varphi_n = \frac{1}{\sqrt{N!}} \sum_{\sigma \in S_N} \text{sign}\,(\sigma)\, \varphi_{\sigma(1)} \otimes \ldots \otimes \varphi_{\sigma(N)}\,,$$

where the sum is over all permutations σ of $\{1,\dots,N\}$, and sign $(\sigma) = \pm 1$ is the sign of the permutation. We consider the variational approximation (1.2) on the manifold

$$\mathcal{M} = \{u \in L^2(\mathbb{R}^{3N}) \,:\, u \neq 0,\; u = a\,\varphi_1 \wedge \ldots \wedge \varphi_N\,,\; a \in \mathbb{C},\; \varphi_n \in L^2(\mathbb{R}^3)\}\,. \tag{3.14}$$

The representation of $u \in \mathcal{M}$ as $u = a\,\varphi_1 \wedge \ldots \wedge \varphi_N$ again is not unique: u remains unaltered under the transformation by any invertible $N \times N$ matrix, $A \in \mathrm{GL}(N)$, by

$$\begin{pmatrix} \varphi_1 \\ \vdots \\ \varphi_N \end{pmatrix} \to A \begin{pmatrix} \varphi_1 \\ \vdots \\ \varphi_N \end{pmatrix}, \qquad a \to \frac{a}{\det(A)}\,.$$

We may therefore choose to work with orthonormal orbitals:

$$\langle \varphi_n \,|\, \varphi_j \rangle = \delta_{nj} \qquad \text{for all } n, j\,. \tag{3.15}$$

In particular, we then have $\|u\| = |a|$.

Tangent Functions. Consider $u = a\,\varphi_1 \wedge \ldots \wedge \varphi_N$ with a of unit modulus and with orthonormal orbitals φ_n. Every tangent function $\dot u \in \mathcal{T}_u\mathcal{M}$ is of the form

$$\dot u = \dot a\,\varphi_1 \wedge \ldots \wedge \varphi_N + a\,\dot\varphi_1 \wedge \varphi_2 \wedge \ldots \wedge \varphi_N + \ldots + a\,\varphi_1 \wedge \ldots \wedge \varphi_{N-1} \wedge \dot\varphi_N \tag{3.16}$$

where $\dot a \in \mathbb{C}$ and $\dot\varphi_n \in L^2$. These turn out to be uniquely determined by $\dot u$ and the fixed $a, \varphi_1, \dots, \varphi_n$ if we impose the gauge condition

$$\langle \varphi_n \,|\, \dot\varphi_j \rangle = 0 \qquad \text{for all } n, j\,. \tag{3.17}$$

Indeed, taking the inner product of both sides of (3.4) with $u = a\,\varphi_1 \wedge \ldots \wedge \varphi_N$ and using (3.15) and (3.17) and $a = 1/\overline{a}$, determines $\dot a$ again as

$$\dot a = \langle u \,|\, \dot u \rangle\, a\,. \tag{3.18}$$

Taking the inner product with the function in which φ_n is replaced by some L^2 function ϑ_n, determines $\dot\varphi_n$ uniquely by the analogue of (3.7), where now simply the wedge product replaces the tensor product:

$$\langle \vartheta_n \,|\, \dot\varphi_n \rangle + \bar a \dot a \, \langle \vartheta_n \,|\, \varphi_n \rangle = \langle a\, \varphi_1 \wedge \ldots \wedge \vartheta_n \wedge \ldots \wedge \varphi_N \,|\, \dot u \rangle \quad \forall \vartheta_n \in L^2 . \tag{3.19}$$

Equations of Motion for the Orbitals. The equations of motion for the orbitals in the variational approximation (1.2) on the Hartree–Fock manifold $\mathcal{M}$ in the weak form therefore still are of the same type as in (3.9), where just $\wedge$ formally replaces $\otimes$. With the constant total energy $E = \langle u \,|\, H \,|\, u \rangle$, we have

$$i\hbar \Big\langle \vartheta_n \,\Big|\, \frac{\partial \varphi_n}{\partial t} \Big\rangle = \langle a\, \varphi_1 \wedge \ldots \wedge \vartheta_n \wedge \ldots \wedge \varphi_N \,|\, Hu \rangle - E \langle \vartheta_n \,|\, \varphi_n \rangle \quad \forall \vartheta_n \in L^2 . \tag{3.20}$$

To proceed further, we now consider a Hamiltonian composed of identical one- and two-body Hamiltonians:

$$H = \sum_{j=1}^{N} \Big(-\frac{\hbar^2}{2m} \Delta_{x_j} + U_j \Big) + \sum_{k<\ell} W_{k\ell} \equiv \sum_{j=1}^{N} S_j + \sum_{k<\ell} W_{k\ell} \tag{3.21}$$

with identical one-body potentials $U_j(x_1, \ldots, x_N) = U(x_j)$ and identical symmetric two-body potentials

$$W_{k\ell}(x_1, \ldots, x_N) = W(x_k, x_\ell) = W(x_\ell, x_k) .$$

The situation of primary interest is that of the electronic Schrödinger equation (2.4), where

$$W(x, y) = \frac{e^2}{|x-y|} \qquad (x, y \in \mathbb{R}^3)$$

is the Coulomb potential of electron-electron interaction, and $U(x)$ describes the Coulomb interaction between an electron at $x \in \mathbb{R}^3$ and all nuclei at fixed positions.

We abbreviate the single-particle operator as $S = -\frac{\hbar^2}{2m} \Delta_x + U$, and write S_j when it is considered as an operator acting on the variable x_j of functions of $(x_1, \ldots, x_N)$.

We return to (3.20) and consider functions $\vartheta_n \in L^2(\mathbb{R}^3)$ that satisfy the orthogonality condition

$$\langle \vartheta_n \,|\, \varphi_j \rangle = 0 \quad \text{for all } n, j . \tag{3.22}$$

Using the definition of the wedge product and the orthogonality relations (3.15) and (3.22) we calculate

$$\langle \varphi_1 \wedge \ldots \wedge \vartheta_n \wedge \ldots \wedge \varphi_N \,|\, S_1 \,|\, \varphi_1 \wedge \ldots \wedge \varphi_N \rangle = \frac{1}{N} \langle \vartheta_n \,|\, S \varphi_n \rangle .$$

Since the same result is obtained for $S_2, \ldots, S_N$, we obtain

$$\Big\langle \varphi_1 \wedge \ldots \wedge \vartheta_n \wedge \ldots \wedge \varphi_N \,\Big|\, \sum_{j=1}^{N} S_j \,\Big|\, \varphi_1 \wedge \ldots \wedge \varphi_N \Big\rangle = \langle \vartheta_n \,|\, S \varphi_n \rangle . \tag{3.23}$$

For the two-body interaction we obtain similarly, using in addition the symmetry of W,

$$\langle\varphi_1 \wedge \ldots \wedge \vartheta_n \wedge \ldots \wedge \varphi_N \,|\, W_{12} \,|\, \varphi_1 \wedge \ldots \wedge \varphi_N\rangle$$
$$= \frac{2}{N(N-1)} \sum_{j \neq n} \Big(\langle \vartheta_n \otimes \varphi_j \,|\, W \,|\, \varphi_n \otimes \varphi_j \rangle - \langle \vartheta_n \otimes \varphi_j \,|\, W \,|\, \varphi_j \otimes \varphi_n \rangle \Big),$$

and the same result for the other $W_{k\ell}$. Hence we have

$$\langle\varphi_1 \wedge \ldots \wedge \vartheta_n \wedge \ldots \wedge \varphi_N \,|\, \sum_{k<\ell} W_{k\ell} \,|\, \varphi_1 \wedge \ldots \wedge \varphi_N\rangle$$
$$= \langle \vartheta_n \,|\, K_n \varphi_n - \sum_{j \neq n} X_{nj} \varphi_j \rangle \tag{3.24}$$

with the *Hartree potential* K_n and the *exchange potentials* X_{nj} given as

$$K_n(x) = \sum_{j \neq n} \int_{\mathbb{R}^3} W(x,y) \, |\varphi_j(y)|^2 \, dy \tag{3.25}$$
$$X_{nj}(x) = \int_{\mathbb{R}^3} W(x,y) \, \overline{\varphi}_j(y) \, \varphi_n(y) \, dy \,. \tag{3.26}$$

Substituting (3.23) and (3.24) into (3.20), we thus obtain, for all $\vartheta_n \in L^2(\mathbb{R}^3)$ satisfying the orthogonality relations (3.22),

$$\Big\langle \vartheta_n \,\Big|\, i\hbar \frac{\partial \varphi_n}{\partial t} - S\varphi_n - K_n \varphi_n + \sum_{j \neq n} X_{nj} \varphi_j \Big\rangle = 0 \,.$$

It follows that the right-hand expression in the inner product is in the linear span of $\varphi_1, \ldots, \varphi_N$. Since adding such a term to $\partial \varphi_n / \partial t$ adds to $\dot{u} = du/dt$ of (3.16) only a scalar multiple of u and hence changes only the scalar phase factor a, the effect of this term is put into a and we set the right-hand expression in the inner product to zero. On multiplying with φ_j and interchanging n and j, we then further obtain

$$\frac{d}{dt} \langle \varphi_n \,|\, \varphi_j \rangle = \Big\langle \varphi_n \,\Big|\, \frac{\partial \varphi_j}{\partial t} \Big\rangle + \overline{\Big\langle \varphi_j \,\Big|\, \frac{\partial \varphi_n}{\partial t} \Big\rangle} = 0 \,,$$

so that the orthonormality relations (3.15) are preserved for all times. Since we know that the variational approximation $u(t)$ conserves the unit norm, the phase factor $a(t)$ then remains of unit modulus. We summarize the result as follows.

Theorem 3.3 (Time-Dependent Hartree–Fock Method, Dirac 1930). *For a Hamiltonian (3.21), the variational approximation (1.2) on the Hartree–Fock manifold (3.14), for initial data* $u(x_1, \ldots, x_N, 0) = \frac{1}{\sqrt{N!}} \det\big(\varphi_n(x_j, 0)\big)_{n,j=1}^N$ *with* $\varphi_n(\cdot, 0)$ *satisfying the orthonormality relations (3.15), is given as*

$$u(x_1, \ldots, x_N, t) = a(t) \, \frac{1}{\sqrt{N!}} \, \det\big(\varphi_n(x_j, t)\big)_{n,j=1}^N \,,$$

where $|a(t)| = 1$ *and* $\varphi_n(x,t)$ *are solutions to the system of partial differential equations*

$$i\hbar \frac{\partial \varphi_n}{\partial t} = -\frac{\hbar^2}{2m} \Delta \varphi_n + U \varphi_n + K_n \varphi_n - \sum_{j \neq n} X_{nj} \varphi_j \,, \tag{3.27}$$

which is nonlinear through the Hartree and exchange potentials given by (3.25) and (3.26). This holds on time intervals $0 \leq t \leq \bar{t}$ *on which a strong solution to this system exists, that is, for* $\varphi_n \in C^1([0,\bar{t}], L^2(\mathbb{R}^3)) \cap C([0,\bar{t}], H^2(\mathbb{R}^3))$. *The orthonormality (3.15) of the orbitals is preserved on the whole time interval.* □

Comparing (3.27) with the Hartree equations (3.12), we note that the only, but essential difference is in the presence of the fermionic exchange terms $X_{nj}\varphi_j$.

Global existence of strong solutions to the equations of motion (3.27) in the case of Coulomb potentials is shown by Chadam & Glassey (1974). The line of their argument runs as follows: first it is shown by Picard iteration that solutions in H^2 exist *locally* in time, where the growth in the H^2 norm is exponential in terms of a bound of the H^1 norm of the solution. Since the H^1 norm of a strong solution can be bounded by the constant total energy $\langle H \rangle$, it follows that the H^1 norm remains in fact bounded for all times and the H^2 norm grows at worst exponentially.

Spin Orbitals. Electrons are distinguished by their spin which can take the two values up ($\uparrow$) and down ($\downarrow$). In a system with K electrons of spin up and $N-K$ electrons of spin down, the separable approximation with the correct antisymmetry properties is

$$u = a\,(\varphi_1 \wedge \ldots \wedge \varphi_K) \otimes (\varphi_{K+1} \wedge \ldots \wedge \varphi_N) \tag{3.28}$$

with $a \in \mathbb{C}$, $\varphi_n \in L^2(\mathbb{R}^3)$. The equations of motion for variational approximations of this type can be derived in the same way as above and turn out to be identical to (3.27) if the interpretation of inner products is modified as follows: we extend each orbital φ_n to a *spin orbital* $\widehat{\varphi}_n = (\varphi_n, s_n)$ with spin $s_n \in \{\uparrow, \downarrow\}$. For any observable A of orbitals we define

$$\langle \widehat{\varphi}_n \,|\, A \,|\, \widehat{\varphi}_j \rangle = \begin{cases} \langle \varphi_n \,|\, A \,|\, \varphi_j \rangle & \text{if } s_n = s_j \,, \\ 0 & \text{else.} \end{cases}$$

With this interpretation of all arising inner products, the equations of motion (3.27) remain valid for the spin orbitals $\widehat{\varphi}_n$, with non-vanishing exchange terms remaining only between spin orbitals of the same spin.

As opposed to the *unrestricted* Hartree-Fock approximation just described, the *restricted* Hartree-Fock method in the case of an even number N of electrons assumes an equal number $N/2$ of electrons with spin up and spin down with the spin orbitals $(\varphi_n, \uparrow)$ and $(\varphi_n, \downarrow)$ for $n = 1, \ldots, N/2$, that is, with the *same* spatial orbital φ_n for both spin up and spin down. The approximation to the wave function is thus chosen of the form

$$u = a\,(\varphi_1 \wedge \ldots \wedge \varphi_{N/2}) \otimes (\varphi_1 \wedge \ldots \wedge \varphi_{N/2}) \tag{3.29}$$

in the restricted Hartree-Fock method. For an initial state of this type, it is seen that this restricted form is preserved for all times in the equations of motion (3.27) of the unrestricted Hartree-Fock method with $N/2$ electrons of spin up and $N/2$ electrons of spin down. Therefore half of the equations can be dropped in this case.

II.3.3 Multi-Configuration Methods (MCTDH, MCTDHF)

Multi-Configurations. We consider again the Schrödinger equation (3.1) for the nuclei, which are supposed to be distinguishable by their different types or by their well-separated positions. It is to be expected, and has found ample confirmation in computations, that a better approximation to the wave function can be obtained by using a linear combination of tensor products instead of just a single tensor product, as is done in the time-dependent Hartree method of Section II.3.1. We therefore consider approximations

$$\psi(x_1,\dots,x_N,t) \approx \sum_{(j_1,\dots,j_N)} a_{j_1,\dots,j_N}(t)\, \varphi_{j_1}^{(1)}(x_1,t)\cdots\varphi_{j_N}^{(N)}(x_N,t)$$
$$\equiv \sum_J a_J(t)\,\Phi_J(x,t)\,. \tag{3.30}$$

Here, the multi-indices $J = (j_1,\dots,j_N)$ vary for $j_n = 1,\dots,r_n$, $n = 1,\dots,N$, the $a_J(t)$ are complex coefficients depending only on t, and the single-particle functions $\varphi_{j_n}^{(n)}(x_n,t)$ depend on the coordinates $x_n \in \mathbb{R}^3$ of particle n and on time t. Alternatively, we might take Hartree products of $3N$ functions depending on $x_n \in \mathbb{R}$.

This is a model reduction analogous to low-rank approximation of matrices, where a large system matrix is replaced by a linear combination of rank-1 matrices $v \otimes w$, or to low-rank approximation of tensors by linear combinations of rank-1 tensors $v_1 \otimes \dots \otimes v_N$.

In the *multi-configuration time-dependent Hartree* (MCTDH) method proposed by Meyer, Manthe & Cederbaum (1990) and developed further as described by Beck, Jäckle, Worth & Meyer (2000), the Dirac–Frenkel time-dependent variational principle (1.2) is used to derive differential equations for the coefficients a_J and the single-particle functions $\varphi_{j_n}^{(n)}$. The MCTDH method determines approximations to the wave function that, for every time t, lie in the set

$$\overline{\mathcal{M}} = \Big\{ u \in L^2(\mathbb{R}^{3N}) : u = \sum_J a_J\, \varphi_{j_1}^{(1)} \otimes \cdots \otimes \varphi_{j_N}^{(N)} \text{ with } a_J \in \mathbb{C},\ \varphi_{j_n}^{(n)} \in L^2(\mathbb{R}^3) \Big\}$$

with multi-indices $J = (j_1,\dots,j_N)$ ranging over $j_n = 1,\dots,r_n$. This set $\overline{\mathcal{M}}$ is not a manifold, but it contains a dense subset $\mathcal{M}$ that is a manifold and is characterized by a full-rank condition to be given below.

The representation of $u \in \overline{\mathcal{M}}$ by a coefficient tensor $A = (a_J)$ and single-particle functions $\Phi = \big(\varphi_{j_n}^{(n)}\big)$ clearly is not unique: the transformation

$$\varphi_{j_n}^{(n)} \to \widehat{\varphi}_{j_n}^{(n)} = \sum_{k_n=1}^{r_n} S_{j_n,k_n}^{(n)} \varphi_{k_n}^{(n)},$$
$$a_J \to \widehat{a}_J = \sum_{i_1=1}^{r_1} \cdots \sum_{i_N=1}^{r_N} a_I (S^{(1)})^{-1}_{i_1,j_1} \cdots (S^{(N)})^{-1}_{i_N,j_N}$$

yields the same function u for any choice of nonsingular matrices $S^{(1)},\dots,S^{(N)}$. We may assume that the orbitals $\varphi_{j_n}^{(n)}$ corresponding to the same particle n are orthonormal:

$$\langle \varphi_{j_n}^{(n)} \,|\, \varphi_{k_n}^{(n)} \rangle = \delta_{j_n,k_n}\,, \qquad j_n, k_n = 1, \dots, r_n, \;\; n = 1, \dots, N\,. \tag{3.31}$$

Tangent Functions. Consider a differentiable path $(A(t), \Phi(t))$ of coefficients and single-particle functions representing a path $u(t)$ on $\overline{\mathcal{M}}$. Then, the derivative $\dot{u}$ is of the form

$$\dot{u} = \sum_J \dot{a}_J \Phi_J + \sum_{n=1}^{N} \sum_{j_n=1}^{r_n} \dot{\varphi}_{j_n}^{(n)} \psi_{j_n}^{(n)} \tag{3.32}$$

with the Hartree products $\Phi_J = \bigotimes_{n=1}^{N} \varphi_{j_n}^{(n)}$ and with the *single-hole functions*

$$\begin{aligned} \psi_{j_n}^{(n)} &= \langle \varphi_{j_n}^{(n)} \,|\, u \rangle^{(n)} \\ &= \sum_{j_1=1}^{r_1} \cdots \sum_{j_{n-1}=1}^{r_{n-1}} \sum_{j_{n+1}=1}^{r_{n+1}} \cdots \sum_{j_N=1}^{r_N} a_{j_1,\dots,j_N} \bigotimes_{k\neq n} \varphi_{j_k}^{(k)}\,, \end{aligned} \tag{3.33}$$

where the superscript (n) on the inner product indicates that the L^2 inner product is taken only with respect to the variable x_n, leaving a function depending on all the other variables x_k with $k \neq n$.

Conversely, the $\dot{a}_J$ are uniquely determined by $\dot{u}$ and (A, Φ) if we impose the orthogonality constraints (which are gauge conditions)

$$\langle \varphi_{j_n}^{(n)} \,|\, \dot{\varphi}_{k_n}^{(n)} \rangle = 0\,, \qquad j_n, k_n = 1, \dots, r_n, \;\; n = 1, \dots, N\,, \tag{3.34}$$

which together with (3.31) implies

$$\dot{a}_J = \langle \Phi_J \,|\, \dot{u} \rangle\,. \tag{3.35}$$

Taking the L^2 inner product of (3.32) with $\psi_{i_n}^{(n)}$ over all variables except x_n, as indicated by the superscript $(\neg n)$, then gives

$$\sum_{j_n=1}^{r_n} \rho_{i_n,j_n}^{(n)} \dot{\varphi}_{j_n}^{(n)} = \Big\langle \psi_{i_n}^{(n)} \,\Big|\, \dot{u} - \sum_J \dot{a}_J \Phi_J \Big\rangle^{(\neg n)} \tag{3.36}$$

with the hermitian, positive semi-definite *density matrices*

$$\rho^{(n)} = \big(\rho_{i_n,j_n}^{(n)}\big)_{i_n,j_n=1}^{r_n} \qquad \text{given by} \qquad \rho_{i_n,j_n}^{(n)} := \langle \psi_{i_n}^{(n)} | \psi_{j_n}^{(n)} \rangle^{(\neg n)}\,. \tag{3.37}$$

The orthonormality relations (3.31) allow us to express the entries of the density matrices in terms of the coefficients a_J:

$$\rho_{i_n,j_n}^{(n)} = \sum_{j_1=1}^{r_1} \cdots \sum_{j_{n-1}=1}^{r_{n-1}} \sum_{j_{n+1}=1}^{r_{n+1}} \cdots \sum_{j_N=1}^{r_N} \bar{a}_{j_1,\dots,j_{n-1},i_n,j_{n+1},\dots,j_N} a_{j_1,\dots,j_N}\,. \tag{3.38}$$

The $\dot{\varphi}_{j_n}^{(n)}$ are thus uniquely determined from (3.36) under the *full-rank condition* that

$$\rho^{(n)} \text{ is an invertible matrix for each } n = 1, \ldots, N. \tag{3.39}$$

(In view of (3.38), a necessary condition for this property is $r_n \le \prod_{k \ne n} r_k$.)

The MCTDH manifold. With the above construction of the $\dot{a}_J$ and $\dot{\varphi}_{j_n}^{(n)}$, one can construct local charts on

$$\begin{aligned}\mathcal{M} \;=\; \Big\{ u \in L^2(\mathbb{R}^{3N}) : \; & u = \sum_J a_J \, \varphi_{j_1}^{(1)} \otimes \cdots \otimes \varphi_{j_N}^{(N)} \text{ with } a_J \in \mathbb{C} \text{ and} \\ & \varphi_{j_n}^{(n)} \in L^2(\mathbb{R}^3) \text{ satisfying the orthonormality condition (3.31)} \\ & \text{and the full-rank condition (3.39)} \Big\}, \end{aligned} \tag{3.40}$$

making this set an infinite-dimensional manifold, for which the tangent space at $u \in \mathcal{M}$ consists of the elements $\dot{u}$ of the form (3.32). We also note that $\|u\|^2 = \sum_J |a_J|^2$.

Equations of Motion for the Multi-Configuration Time-Dependent Hartree Method. The MCTDH method uses the time-dependent variational principle (1.2) on this approximation manifold $\mathcal{M}$. The equations of motion are thus obtained by substituting $\frac{1}{i\hbar} H u$ for $\dot{u}$ in (3.35) and (3.36), and so we have the following result.

Theorem 3.4 (MCTDH Method; Meyer, Manthe & Cederbaum 1990). *The variational approximation on the MCTDH manifold (3.40) is given by (3.30), where the coefficients and single-particle functions are solutions to the system of coupled ordinary and partial differential equations*

$$i\hbar \frac{da_J}{dt} = \sum_K \langle \Phi_J | H | \Phi_K \rangle \, a_K \,, \qquad \forall J = (j_1, \ldots, j_N), \tag{3.41}$$

$$i\hbar \frac{\partial \varphi_{j_n}^{(n)}}{\partial t} = (I - P^{(n)}) \sum_{k_n=1}^{r_n} \sum_{l_n=1}^{r_n} (\rho^{(n)})^{-1}_{j_n,k_n} \langle \psi_{k_n}^{(n)} \,|\, H \,|\, \psi_{l_n}^{(n)} \rangle^{(\neg n)} \, \varphi_{l_n}^{(n)}, \tag{3.42}$$

$$j_n = 1, \ldots, r_n, \;\; n = 1, \ldots, N.$$

This holds on every time interval on which a strong solution to these equations exists. Here, the Hartree products Φ_J, the single-hole functions $\psi_{j_n}^{(n)}$, and the density matrices $\rho^{(n)}$ are defined in (3.30), (3.33), and (3.38), respectively. The superscript $(\neg n)$ indicates that the inner product is over all variables except x_n, and $P^{(n)}$ is the orthogonal projector onto the linear span of $\varphi_1^{(n)}, \ldots, \varphi_{r_n}^{(n)}$. □

We note that the projector $P^{(n)}$ is given as $P^{(n)}\vartheta = \sum_{j_n=1}^{r_n} \varphi_{j_n}^{(n)} \langle \varphi_{j_n}^{(n)} \,|\, \vartheta \rangle^{(n)}$, with the inner product over the variable x_n.

For a smooth bounded potential with bounded derivatives, it is shown by Koch & Lubich (2007) that a strong solution $\varphi_{j_n}^{(n)} \in C^1([0, \bar{t}), L^2(\mathbb{R}^3)) \cap C([0, \bar{t}), H^2(\mathbb{R}^3))$ to the MCTDH equations exists either globally for all times or up to a time $\bar{t}$ where a density matrix $\rho^{(n)}$ becomes singular.

At a singularity of a density matrix $\rho^{(n)}$, the equations of motion break down. To avoid such problems in computations, the density matrices are usually regularized to $\rho^{(n)} + \mu I$ with a small parameter μ. Although such regularized solutions exist for all times, a near-singularity can still cause numerical problems, in particular in the step size selection of a time integration method.

The MCTDH method has been used successfully for accurately computing the quantum dynamics of small molecules in a variety of chemical situations such as photodissociation and reactive scattering, for problems involving 6 to 24 nuclear degrees of freedom and one or several electronic states; see, e.g., Raab, Worth, Meyer & Cederbaum (1999).

The complexity of the method grows exponentially with the number of particles: there are r^N coefficients a_J if $r_n = r$ orbitals are taken for each particle. Several variants and extensions of the MCTDH method have been designed for the computational treatment of larger systems, such as the coupling with Gaussian wavepackets for secondary modes (Burghardt, Meyer & Cederbaum 1999) and the *hierarchical, cascadic* or *multilayer* versions of MCTDH (Beck, Jäckle, Worth & Meyer 2000, Wang & Thoss 2003) with which particular systems of up to 500 degrees of freedom have been treated.

Hierarchical MCTDH Method. Considering for simplicity a system with $N = 2^L$ particles, the binary cascadic MCTDH method determines an approximation to the wave function in the form

$$u = \sum_{j,k=1}^{r} a_{jk}\, \varphi_j^{(0)} \otimes \varphi_k^{(1)}$$

where, for a binary number $B = (b_1, \ldots, b_\ell)$ with $b_m \in \{0,1\}$ and $\ell < L$, we set recursively

$$\varphi_i^B = \sum_{j,k=1}^{r} a_{i,jk}^B\, \varphi_j^{(B,0)} \otimes \varphi_k^{(B,1)},$$

and for $\ell = L$ we have the single-particle functions. The variational approximation u is thus built up from a binary tree, with the single-particle functions sitting at the end of the branches. This approach uses only $r^3 N$ instead of r^N coefficients.

The orthogonality relations (3.31) and (3.34) can now be imposed on each level: at the final level for the single-particle functions and at the other levels by

$$\langle \varphi_\ell^B \,|\, \varphi_i^B \rangle = \sum_{j,k=1}^{r} \overline{a_{\ell,jk}^B}\, a_{i,jk}^B = \delta_{i\ell}$$

$$\langle \varphi_\ell^B \,|\, \dot\varphi_i^B \rangle = \sum_{j,k=1}^{r} \overline{a_{\ell,jk}^B}\, \dot a_{i,jk}^B = 0\,.$$

The derivation of the equations of motion is then analogous to that of the MCTDH method given above, with recurrences climbing up and down the tree for the computation of the required inner products.

Multi-Configuration Time-Dependent Hartree-Fock Method (MCTDHF). For electron dynamics, a multi-configuration extension of the TDHF method is obtained by using the time-dependent variational principle for approximations of the form

$$u = \sum_{1 \le j_1 < \ldots < j_N \le K} c_{j_1 \ldots j_N} \, \varphi_{j_1} \wedge \ldots \wedge \varphi_{j_N} \tag{3.43}$$

with $K > N$. The sum is over all $\binom{K}{N}$ subsets with N elements of $\{1, \ldots, K\}$. The equations of motion of the MCTDHF method for a Hamiltonian (3.21) are those of the MCTDH method with $\varphi_j^{(n)} = \varphi_j$ independent of n and with an antisymmetric tensor: for every multi-index $J = (j_1, \ldots, j_N)$ and permutation $\sigma \in S_N$ and with $\sigma(J) = (j_{\sigma(1)}, \ldots, j_{\sigma(N)})$,

$$a_{\sigma(J)} = \operatorname{sign}(\sigma) \, a_J \, .$$

We refer to Zanghellini, Kitzler, Fabian, Brabec & Scrinzi (2003) and Koch, Kreuzer & Scrinzi (2006) for uses and properties of the MCTDHF approach.

No Theoretical Approximation Estimates. While the neighbouring sections close with theoretical results on the approximation error, apparently no such results are available for the methods considered in this section. One might hope that the multi-configuration methods converge to the exact wave function as the number of configurations is increased to infinity, but to date no such result exists. One obstacle to such a convergence result is the fact that the density matrices $\rho^{(n)}$ become more and more ill-conditioned as more nearly irrelevant configurations are included. Another difficulty lies in the time-dependent orbitals whose approximation properties are not under control. In Sect. II.6 we show, however, that for a *fixed* number of configurations, the variational approximation is quasi-optimal in the sense that its error – on sufficiently short time intervals – is bounded in terms of the error of the best approximation to the wave function by the given number of configurations.

Notwithstanding the deficiencies in theory, the methods considered in this section have proven their value in computations on realistic chemical and physical systems — the tool apparently works.

II.4 Parametrized Wave Functions: Gaussian Wave Packets

A further modelling or approximation step consists in replacing the wave function by a function that depends only on a finite number of real or complex parameters. The time-dependent variational principle then yields evolution equations for these parameters that retain a Hamiltonian character, albeit with a non-canonical Poisson bracket. Gaussian wave packets parametrized by position, momentum, complex width and phase are a prominent example. In the classical limit, their variational equations of motion for position and momentum yield the Newtonian equations of classical molecular dynamics.

II.4.1 Variational Gaussian Wave-Packet Dynamics

We consider a Schrödinger equation in semi-classical scaling, for $x \in \mathbb{R}^d$,

$$i\varepsilon \frac{\partial \psi}{\partial t} = H\psi\,, \qquad H = H^\varepsilon = -\frac{\varepsilon^2}{2}\Delta + V\,, \tag{4.1}$$

with a small positive parameter $\varepsilon \ll 1$ (formally in place of $\hbar$, see Sect. II.2.3) and a potential V. The typical situation of (4.1) is the time-dependent Born-Oppenheimer approximation for the motion of nuclei, where ε^2 represents a mass ratio of nuclei and electrons.

As proposed by Heller (1976), the variational approximation of (4.1) can be done by complex Gaussians of the type

$$\psi(x,t) \approx u(x,t) = \exp\Bigl(\frac{i}{\varepsilon}\Bigl(\frac{1}{2}(x-q(t))^T C(t)(x-q(t)) + p(t)\cdot(x-q(t)) + \zeta(t)\Bigr)\Bigr), \tag{4.2}$$

where $q(t) \in \mathbb{R}^d$ is the position average and $p(t) \in \mathbb{R}^d$ is the momentum average of the wave packet. The matrix $C(t) \in \mathbb{C}^{d\times d}$ is a complex symmetric width matrix with positive definite imaginary part, possibly further restricted to a diagonal matrix or just a multiple of the identity, $c(t)I_d$ with complex $c(t)$. Finally, $\zeta(t) \in \mathbb{C}$ is a phase and normalization parameter.

The choice of Gaussians appears attractive because the exact wave function retains the form of a multidimensional Gaussian for all times in the case of a quadratic potential, even for a time-dependent quadratic potential. This useful fact follows from the observation that Hu then is in the tangent space at u, and therefore the variational approximation and the exact wave function coincide. For a narrow wave packet, of width $\sim \varepsilon^{1/2}$ in (4.2), a smooth potential appears locally approximately quadratic, and we may then expect good approximation by Gaussians, as will be made more precise in Sect. II.4.4 in an argument based on the error bound (1.11).

The equations of motion for the parameters read as follows (Heller 1976, Coalson & Karplus 1990): with $\langle A\rangle = \langle u\,|\,A\,|\,u\rangle$ denoting the average of an observable A in the Gaussian state u of unit L^2 norm, we have classically-looking equations for position and momentum, with the average of the gradient ∇V of the potential,

$$\begin{aligned} \dot q &= p \\ \dot p &= -\langle \nabla V\rangle\,. \end{aligned} \tag{4.3}$$

For the width matrix C and the complex phase ζ we have, with the Hessian $\nabla^2 V$ and with tr denoting the trace of a matrix,

$$\dot C = -C^2 - \langle \nabla^2 V\rangle \tag{4.4}$$

$$\dot\zeta = \frac{1}{2}|p|^2 - \langle V\rangle + \frac{i\varepsilon}{2}\,\mathrm{tr}\, C + \frac{\varepsilon}{4}\bigl\langle \mathrm{tr}\,\bigl((\mathrm{Im}\, C)^{-1}\nabla^2 V\bigr)\bigr\rangle\,. \tag{4.5}$$

When C is restricted to diagonal matrices, then the diagonal part is to be taken on the right-hand side of the differential equation for C. When $C = cI$ is restricted to a multiple of the identity (spherical Gaussians), then the differential equation for c is obtained by taking the trace on both sides of the differential equation for C. If the width matrix is taken constant (frozen Gaussians, Heller 1981), then the equation for C is discarded, and only the equations for position and momentum and phase remain.

The more general situation of a Hamiltonian

$$H = -\sum_{n=1}^{N} \frac{\varepsilon^2}{2m_n} \Delta_{x_n} + V$$

with different mass parameters m_n, collected in the diagonal mass matrix $M = \mathrm{diag}(m_n)$, is readily reduced to (4.1). This is done by transforming to variables $\widehat{x} = M^{1/2}x$ and to

$$\widehat{q} = M^{1/2}q\,, \quad \widehat{p} = M^{-1/2}p\,, \quad \widehat{C} = M^{-1/2}CM^{-1/2}\,, \quad \widehat{\zeta} = \zeta\,,$$

which again evolve according to the differential equations (4.3)–(4.5).

As $\varepsilon \to 0$, the Gaussians (4.2) become narrower and increasingly concentrated at q, and we have $\langle \nabla V \rangle \to \nabla V(q)$ for a Gaussian of unit L^2 norm. Hence the equations for position q and momentum p become the

classical equations of motion in the limit $\varepsilon \to 0$.

The differential equations (4.3)–(4.5) are a regular perturbation to the equations for $\varepsilon = 0$: letting $\varepsilon \to 0$ gives a well-defined limit on the right-hand side. They are no longer a singularly perturbed system as (4.1) is. In contrast to the Gaussian wave packet, the time-dependent parameters are not highly oscillatory functions.

We shall give a derivation of the equations of motion (4.3)–(4.5) that highlights their mathematical structure as a non-canonical Hamiltonian system (or a Poisson system in another terminology). We first study the structure of the variational equations of motion in coordinates on an approximation manifold in a general setting and then return to the particular case of Gaussian wave packets. The presentation in this section essentially follows Faou & Lubich (2006).

II.4.2 Non-Canonical Hamilton Equations in Coordinates

Canonical Poisson Structure of the Schrödinger Equation. We split $\psi \in L^2(\mathbb{R}^d, \mathbb{C})$ into the real and imaginary parts $\psi = v + iw$. The functions v and w are thus functions in the *real* Hilbert space $L^2(\mathbb{R}^d, \mathbb{R})$. We denote the complex inner product by $\langle \cdot \,|\, \cdot \rangle$ and the real inner product by $(\cdot \,|\, \cdot)$.

As the Hamiltonian H is a real operator, the Schrödinger equation (4.1) can be written

$$\begin{aligned} \varepsilon \dot{v} &= Hw, \\ \varepsilon \dot{w} &= -Hv\,. \end{aligned} \tag{4.6}$$

With the canonical structure matrix

$$J = \begin{pmatrix} 0 & -1 \\ 1 & 0 \end{pmatrix}$$

and the Hamiltonian function

$$H(v,w) = \langle \psi \,|\, H \,|\, \psi \rangle = (v \,|\, Hv) + (w \,|\, Hw)$$

for $\psi = v + iw$ (we use the same symbol H as for the operator), this becomes the canonical Hamiltonian system

$$\begin{pmatrix} \dot v \\ \dot w \end{pmatrix} = \frac{1}{2\varepsilon} J^{-1} \nabla H(v,w) \,. \tag{4.7}$$

We note that the real multiplication with J corresponds to the complex multiplication with the imaginary unit i.

As in Theorem 1.2, the flow of this system preserves the canonical symplectic two-form

$$\omega(\xi,\eta) = 2\varepsilon\, (\xi \,|\, J\eta) = (\xi_2 \,|\, \eta_1) - (\xi_1 \,|\, \eta_2), \qquad \xi,\eta \in L^2(\mathbb{R}^d,\mathbb{R})^2. \tag{4.8}$$

The associated Poisson bracket is

$$\{F,G\}_{\mathrm{can}} = \frac{1}{2\varepsilon} (\nabla F \,|\, J^{-1} \nabla G) \tag{4.9}$$

for functions $F, G : H^1(\mathbb{R}^d,\mathbb{R})^2 \to \mathbb{R}$. We have

$$\frac{d}{dt} F(v(t),w(t)) = \{F,H\}_{\mathrm{can}}(v(t),w(t))$$

along the solutions of (4.7).

Poisson Structure of Variational Approximations. We consider a finite-dimensional smooth submanifold $\mathcal{M}$ (of dimension m) of the complex Hilbert space $L^2(\mathbb{R}^d,\mathbb{C})$ with the property (1.3), i.e., with $\mu \in \mathcal{T}_u\mathcal{M}$ also $i\mu \in \mathcal{T}_u\mathcal{M}$ at every $u \in \mathcal{M}$.

Taking the imaginary part in the Dirac–Frenkel time-dependent variational principle (1.2) on $\mathcal{M}$ yields, upon identifying $u = v + iw$ with the real pair $u = (v,w)^T$,

$$\big(\mu \,|\, 2\varepsilon J \dot u - \nabla_u H(u)\big) = 0 \quad \text{for all} \quad \mu \in \mathcal{T}_u\mathcal{M} \,. \tag{4.10}$$

We choose (local) coordinates on $\mathcal{M}$ so that we have a smooth parametrization

$$u = \chi(y)$$

of $\mathcal{M}$, for y in an open subset of $\mathbb{R}^m$. We denote the derivative $X_{\mathbb{C}}(y) = d\chi(y) = V(y) + iW(y)$ or in the real setting as $X = \begin{pmatrix} V \\ W \end{pmatrix}$, which is of full rank for a coordinate map χ. We denote by X^T the adjoint of X with respect to the real inner product $(\cdot\,|\,\cdot)$.

Since $\dot{u} = X(y)\dot{y}$ and the tangent vectors in $\mathcal{T}_u\mathcal{M}$ are given as $\mu = X(y)\eta$ with arbitrary $\eta \in \mathbb{R}^m$, we obtain from (4.10) the differential equation in $\mathbb{R}^m$,

$$2\varepsilon X(y)^T JX(y)\,\dot{y} = X(y)^T \nabla_u H(\chi(y))\,. \tag{4.11}$$

With $X_{\mathbb{C}}^*$ denoting the adjoint of $X_{\mathbb{C}}$ with respect to the complex inner product $\langle \cdot \mid \cdot \rangle$, we note $X_{\mathbb{C}}^* X_{\mathbb{C}} = (V^TV + W^TW) + i(V^TW - W^TV) = X^TX - iX^TJX$ and hence

$$X^TJX = -\mathrm{Im}\, X_{\mathbb{C}}^* X_{\mathbb{C}}. \tag{4.12}$$

This skew-symmetric matrix is invertible, as the following lemma shows.

Lemma 4.1. *If $\mathcal{T}_u\mathcal{M}$ is a complex linear space for every $u \in \mathcal{M}$, then*

$$X(y)^T JX(y) \ \text{ is invertible for all } y.$$

Proof. We fix $u = \chi(y) \in \mathcal{M}$ and omit the argument y in the following. Since $\mathcal{T}_u\mathcal{M} = \mathrm{Range}(X_{\mathbb{C}})$ is complex linear by assumption, there exists a real linear mapping $L : \mathbb{R}^m \to \mathbb{R}^m$ such that $iX_{\mathbb{C}}\eta = X_{\mathbb{C}}L\eta$ for all $\eta \in \mathbb{R}^m$. This implies

$$JX = XL \quad \text{and} \quad L^2 = -\mathrm{Id}$$

and hence $X^TJX = X^TXL$, which is invertible, since X is of full rank. □

We denote the inverse, which is again skew-symmetric, by

$$B(y) = \frac{1}{2\varepsilon}\left(X(y)^T JX(y)\right)^{-1}. \tag{4.13}$$

Introducing the Hamiltonian function on the manifold $\mathcal{M}$ in the coordinates y as

$$K(y) = H(\chi(y)), \tag{4.14}$$

we note $X(y)^T \nabla_u H(\chi(y)) = \nabla_y K(y)$ in (4.11). We then have the following result.

Theorem 4.2 (Variational Equations of Motion in Coordinates). *The differential equations of the variational approximation in coordinates read*

$$\dot{y} = B(y)\nabla_y K(y)\,. \tag{4.15}$$

This is a non-degenerate Poisson system, i.e., the structure matrix $B(y)$ is invertible and generates a bracket

$$\{F, G\}(y) = \nabla F(y)^T B(y) \nabla G(y) \tag{4.16}$$

on smooth real-valued functions F, G, which is antisymmetric ($\{G, F\} = -\{F, G\}$) and satisfies the Jacobi identity ($\{E, \{F, G\}\} + \{F, \{G, E\}\} + \{G, \{E, F\}\} = 0$) and the Leibniz rule ($\{E \cdot F, G\} = E \cdot \{F, G\} + F \cdot \{E, G\}$).

Proof. By (4.11) and the definitions of $B(y)$ and $K(y)$, we have (4.15). It remains to prove the properties of the bracket. Since ε plays no role here, we let $B(y) = \big(X(y)^T J X(y)\big)^{-1}$ (without the factor $\frac{1}{2\varepsilon}$) in this proof. For points $u \in \mathcal{M}$ we introduce the symplectic projector $\Pi(u)$ from the Hilbert space $\mathcal{H} = L^2(\mathbb{R}^d, \mathbb{R})^2$ onto the tangent space $\mathcal{T}_u\mathcal{M}$ as

$$\Pi(u) = X(y)B(y)X(y)^T J, \qquad u = \chi(y) \in \mathcal{M}\,,$$

From the induced decomposition $\mathcal{H} = \Pi(u)\mathcal{H} \oplus \big(I - \Pi(u)\big)\mathcal{H}$ we obtain, by the implicit function theorem, a corresponding splitting in a neighbourhood of the manifold $\mathcal{M}$ in $\mathcal{H}$,

$$\psi = u + v \quad \text{with } u \in \mathcal{M},\ \Pi(u)v = 0\,.$$

This permits us to extend functions F to a neighbourhood of $\mathcal{M}$ by setting

$$\widehat{F}(\psi) = F(y) \quad \text{for } \psi = u + v \text{ with } u = \chi(y),\ \Pi(u)v = 0\,.$$

We then have for the derivative $d\widehat{F}(u) = d\widehat{F}(u)\Pi(u)$ and hence for its adjoint, the gradient, $\nabla\widehat{F}(u) = \Pi(u)^T\nabla\widehat{F}(u)$. Moreover, $\nabla F(y) = X(y)^T\nabla\widehat{F}(u)$ for $u = \chi(y)$. For the canonical bracket this gives, at $u = \chi(y)$,

$$\begin{aligned}\{\widehat{F}, \widehat{G}\}_{\mathrm{can}}(u) &= \nabla\widehat{F}(u)^T\Pi(u)J^{-1}\Pi(u)^T\nabla\widehat{G}(u)\\ &= \nabla F(y)^T B(y)\nabla G(y) = \{F, G\}(y)\,.\end{aligned}$$

Therefore the stated properties follow from the corresponding properties of the canonical bracket. □

We note that along solutions $y(t)$ of (4.15) we have, for real-valued functions F,

$$\frac{d}{dt}F(y(t)) = \{F, K\}(y(t)).$$

More on Poisson systems can be found in Hairer, Lubich & Wanner (2006), Chap. VII.2, and Marsden & Ratiu (1999), Chap. 8.5. In particular, the flow map $\phi_t : y(0) \mapsto y(t)$ is a *Poisson map*, that is, it preserves the Poisson bracket as

$$\{F \circ \phi_t, G \circ \phi_t\} = \{F, G\} \circ \phi_t \qquad \forall\, F, G\,.$$

The property of being a Poisson map in the coordinates can be translated to be an equivalent formulation of the symplecticity of the flow on the manifold $\mathcal{M}$ as stated by Theorem 1.2.

II.4.3 Poisson Structure of Gaussian Wave-Packet Dynamics

The variational Gaussian wave-packet dynamics (4.3)–(4.5) is obtained by choosing the manifold $\mathcal{M}$ as consisting of complex Gaussians (4.2). For ease of presentation we give the derivation for spherical Gaussians, where $C = cI_d$ with a complex $c = \alpha + i\beta$ with $\beta > 0$, and I_d is the d-dimensional identity. We write the complex phase as $\zeta = \gamma + i\delta$. We then have the approximation manifold

$$\mathcal{M} = \{u = \chi(y) \in L^2(\mathbb{R}^d) : y = (p, q, \alpha, \beta, \gamma, \delta) \in \mathbb{R}^{2d+4} \text{ with } \beta > 0\} \tag{4.17}$$

with

$$\big(\chi(y)\big)(x) = \exp\Big(\frac{i}{\varepsilon}\big((\alpha + i\beta)\,|x - q|^2 + p \cdot (x - q) + \gamma + i\delta\big)\Big). \tag{4.18}$$

The tangent space $\mathcal{T}_u\mathcal{M} \subset L^2(\mathbb{R}^d)$ at a given point $u = \chi(y) \in \mathcal{M}$ is $(2d + 4)$-dimensional and is made of the elements of $L^2(\mathbb{R}^d)$ written as

$$\frac{i}{\varepsilon}\Big((A + iB)\,|x - q|^2 + (P - 2(\alpha + i\beta)Q) \cdot (x - q) - p \cdot Q + C + iD\Big)u \tag{4.19}$$

with arbitrary $(P, Q, A, B, C, D)^T \in \mathbb{R}^{2d+4}$. The tangent space $\mathcal{T}_u\mathcal{M}$ is indeed complex linear (note $\beta > 0$). Moreover, we have $u \in \mathcal{T}_u\mathcal{M}$, and hence Theorem 1.4 shows the preservation of the squared L^2 norm of $u = \chi(y)$, which is given by

$$N(y) = \|\chi(y)\|^2 = \exp\left(-\frac{2\delta}{\varepsilon}\right)\left(\frac{\pi\varepsilon}{2\beta}\right)^{d/2}. \tag{4.20}$$

We then have the following result.

Theorem 4.3 (Gaussian Wave-Packet Dynamics as a Poisson System). *The variational approximation on the Gaussian wave-packet manifold $\mathcal{M}$ of (4.17)–(4.18) yields the Poisson system*

$$\dot{y} = B(y)\nabla_y K(y) \tag{4.21}$$

where, for $y = (p, q, \alpha, \beta, \gamma, \delta) \in \mathbb{R}^{2d+4}$ with $\beta > 0$,

$$B(y) = \frac{1}{N(y)}\begin{pmatrix} 0 & -I_d & 0 & 0 & -p & 0 \\ I_d & 0 & 0 & 0 & 0 & 0 \\ 0 & 0 & 0 & \frac{4\beta^2}{\varepsilon d} & 0 & -\beta \\ 0 & 0 & -\frac{4\beta^2}{\varepsilon d} & 0 & \beta & 0 \\ p^T & 0 & 0 & -\beta & 0 & \frac{d+2}{4}\varepsilon \\ 0 & 0 & \beta & 0 & -\frac{d+2}{4}\varepsilon & 0 \end{pmatrix} \tag{4.22}$$

defines a Poisson structure, and for $u = \chi(y)$,

$$K(y) = \langle\, u \,|\, H \,|\, u \rangle = K_T(y) + K_V(y) \tag{4.23}$$

is the total energy, with kinetic and potential parts

$$K_T(y) = N(y)\left(\frac{|p|^2}{2} + \frac{\varepsilon d}{2}\frac{\alpha^2+\beta^2}{\beta}\right) = \left\langle u \,\Big|\, -\frac{\varepsilon^2}{2}\Delta \,\Big|\, u \right\rangle$$

and

$$K_V(y) = \int_{\mathbb{R}^d} V(x)\exp\Big(-\frac{2}{\varepsilon}(\beta|x-q|^2+\delta)\Big)\,dx = \langle\, u \,|\, V \,|\, u \rangle.$$

Both $K(y)$ and $N(y)$ are conserved quantities of the system.

Proof. By (4.18), the derivative $X_{\mathbb{C}}(y) = d\chi(y) = \big(\frac{\partial u}{\partial p}, \frac{\partial u}{\partial q}, \frac{\partial u}{\partial \alpha}, \frac{\partial u}{\partial \beta}, \frac{\partial u}{\partial \gamma}, \frac{\partial u}{\partial \delta}\big)$ for $u = \chi(y)$ is written

$$X_{\mathbb{C}}(y) = \frac{i}{\varepsilon}\,\big(x-q\,,\, -2(\alpha+i\beta)(x-q)-p\,,\, |x-q|^2\,,\, i|x-q|^2\,,\, 1\,,\, i\big)\,u\,.$$

Calculating the Gaussian integrals, we obtain from (4.12) that

$$2\varepsilon X^T(y)JX(y) = N(y)\begin{pmatrix} 0 & I_d & 0 & 0 & 0 & 0 \\ -I_d & 0 & 0 & \frac{dp}{2\beta} & 0 & \frac{2p}{\varepsilon} \\ 0 & 0 & 0 & -\frac{\varepsilon d(d+2)}{8\beta^2} & 0 & -\frac{d}{2\beta} \\ 0 & -\frac{dp^T}{2\beta} & \frac{\varepsilon d(d+2)}{8\beta^2} & 0 & \frac{d}{2\beta} & 0 \\ 0 & 0 & 0 & -\frac{d}{2\beta} & 0 & -\frac{2}{\varepsilon} \\ 0 & -\frac{2p^T}{\varepsilon} & \frac{d}{2\beta} & 0 & \frac{2}{\varepsilon} & 0 \end{pmatrix}.$$

The inverse of this matrix can be computed explicitly to give the above matrix $B(y)$. Theorem 4.2 then yields the Poisson system, and Theorems 1.1 and 1.4 give the conservation of energy and norm. □

II.4.4 Approximation Error

From the error bound (1.11) we derive the following result, which is closely related to a result by Hagedorn (1980) on non-variational Gaussian wave packets.

Theorem 4.4 (Error Bound for Variational Gaussian Wave Packets). *Consider the variational multidimensional Gaussian wave packet approximation (4.3)–(4.5). Assume that the smallest eigenvalue of the width matrix* $\operatorname{Im} C(t)$ *is bounded from below by a constant $\rho > 0$. Assume that the potential V is three-times continuously differentiable with a bounded third derivative. Then, the error between the Gaussian wave packet $u(t)$ and the exact wave function $\psi(t)$ with Gaussian initial data $\psi(0) = u(0)$ is bounded in the L^2 norm by*

$$\|u(t) - \psi(t)\| \le c\,t\,\varepsilon^{1/2}\,,$$

where c depends only on ρ and the bound of $\partial^3 V$.

Proof. In view of the error bound of Theorem 1.5, we estimate the distance of $\frac{1}{i\varepsilon}Hu(t)$ to the tangent space $\mathcal{T}_{u(t)}\mathcal{M}$. We split the potential into the quadratic Taylor polynomial $U_{q(t)}$ at the current position $q(t)$ and the non-quadratic remainder $W_{q(t)}$,

$$V = U_{q(t)} + W_{q(t)} ,$$

where we note $|W_q(x)| \le \frac{1}{3!} B_3 \, |x-q|^3$ with a bound B_3 of $\partial^3 V$. Since both Δu and $U_q u$ are in the tangent space $\mathcal{T}_u\mathcal{M}$ given by (4.19), we have

$$\text{dist}\left(\frac{1}{i\varepsilon}Hu, \mathcal{T}_u\mathcal{M}\right) = \text{dist}\left(\frac{1}{i\varepsilon}W_q u, \mathcal{T}_u\mathcal{M}\right) \le \left\| \frac{1}{\varepsilon} W_q u \right\| .$$

With the above bound for W_q and the condition on the width matrix we obtain, for a Gaussian state u of unit L^2 norm,

$$\|W_q u\| \le c_1 \left(\varepsilon^{-d/2} \int_{\mathbb{R}^d} e^{-2\rho|x-q|^2/\varepsilon} \, |x-q|^6 \, dx \right)^{1/2} \le c\,\varepsilon^{3/2} ,$$

and hence the result follows with Theorem 1.5. □

As is clear from the proof, the global boundedness of $\partial^3 V$ can be weakened to a bound in a neighbourhood of the positions $q(t)$ and exponential growth outside this region.

We remark that an analogous result does not hold for Gaussian wave packets where the width matrix is restricted to a diagonal matrix.

Though the above result is asymptotically comforting, it must be noted that for realistic values of $\varepsilon \approx 10^{-2}$, a result with a predicted error of $\varepsilon^{1/2}$ cannot necessarily be considered accurate. We will turn to more accurate semi-classical methods briefly in the next section and in more detail in Chapter V.

II.5 Mixed Models, Quantum-Classical Models

There are numerous possibilities for extensions and combinations of the models described in the foregoing sections. For example, within an MCTDH framework, for some parts of the system the single-particle functions might be chosen as Gaussians, while they are left of a general form for other particles (Burghardt, Meyer & Cederbaum, 1999). Considering the Gaussians of frozen width in such a model and passing to the classical limit $\varepsilon \to 0$ in the equations of motions for positions and momenta then yields equations of motion where most particles are described classically while some are treated quantum-mechanically. For example, this is desired for studying proton transfer in a critical region of a molecule, or more generally for describing a quantum subsystem in a classical bath.

II.5.1 Mean-Field Quantum-Classical Model

Among the various possible mixed quantum-classical models, we now describe the conceptually simplest one which has found widespread use in computations, in spite of its known flaws. Consider a system of light and heavy particles (e.g., protons and the other, heavier nuclei in a molecule), where one would like to describe the light particles quantum-mechanically and the heavy particles classically. Let x and y denote the position coordinates of heavy and light particles, respectively. We consider the Schrödinger equation with the Hamiltonian $H = -\frac{\varepsilon^2}{2}\Delta_x - \frac{1}{2}\Delta_y + V(x,y)$, where ε^2 is the mass ratio as in Section II.2.3. We start from a time-dependent Hartree approximation to the full wave function $\Psi(x,y,t)$

$$\Psi(x,y,t) \approx \gamma(x,t)\,\psi(y,t)\,,$$

where we restrict $\gamma(x,t)$ further to take the form of a frozen Gaussian at variable position $q(t)$ and with variable momentum $p(t)$. When we write down the equations of motion for the corresponding variational approximation and let the width of the Gaussians tend to zero, so that averages over x are replaced by evaluations at the position $q(t)$, then we obtain the following coupled system of classical and quantum equations where the classical particles are driven by the mean-field potential of the quantum particles, the wave function of which is determined by a Schrödinger equation with a potential evaluated at the current classical position:

$$\begin{aligned} \dot q &= p \\ \dot p &= -\nabla_q \langle \psi \,|\, V(q,\cdot) \,|\, \psi\rangle \\ i\varepsilon\,\frac{\partial \psi}{\partial t} &= -\frac{1}{2}\Delta\psi + V(q,\cdot)\psi\,. \end{aligned} \tag{5.1}$$

While this appears as an attractive model at first sight, its mean-field character is flawed. The problem becomes clear by the following argument: Suppose we start with an initial wave function

$$\Psi(x,y,0) = \alpha_1\gamma_1^0(x)\Phi_1(x,y) + \alpha_2\gamma_2^0(x)\Phi_2(x,y)\,,$$

where $\Phi_j(x,\cdot)$ are eigenfunctions of $H_e(x) = -\frac{1}{2}\Delta + V(x,\cdot)$ to well-separated eigenvalues $E_j(x)$, of unit L_y^2 norm, and γ_j^0 are complex Gaussians of width $\sim \varepsilon^{1/2}$ and unit L_x^2 norm. The coefficients should satisfy $|\alpha_1|^2 + |\alpha_2|^2 = 1$ so that Ψ is of unit $L_{x,y}^2$ norm. Using first Theorem 2.1 and then Theorem 4.4 shows that for times $t \sim 1$ the exact wave function $\Psi_j(x,y,t)$ with initial data $\gamma_j^0(x)\Phi_j(x,y)$ is approximately, up to an error of order $\varepsilon^{1/2}$,

$$\Psi_j(x,y,t) \approx \gamma_j(x,t)\Phi_j(x,y)\,,$$

where $\gamma_j(x,t)$ is a Gaussian located at a position $q_j(t)$ that follows classical equations of motion

$$\dot q_j = p_j, \quad \dot p_j = -\nabla_q E_j(q_j)\,. \tag{5.2}$$

The total wave function $\Psi = \alpha_1\Psi_1 + \alpha_2\Psi_2$ is thus approximately equal to

$$\Psi(x,y,t) \approx \alpha_1\gamma_1(x,t)\Phi_1(x,y) + \alpha_2\gamma_2(x,t)\Phi_2(x,y)\,.$$

On the other hand, in (5.1) the time-adiabatic theorem mentioned after Theorem 2.1 and applied with the time-dependent Hamiltonian $H_*(t) = -\frac{1}{2}\Delta + V(q(t),\cdot)$, yields that for times $t \sim 1$,

$$\psi(y,t) \approx \beta_1(t)\,\Phi_1(q(t),y) + \beta_2(t)\,\Phi_2(q(t),y)$$

with coefficients satisfying $|\beta_j(t)| = |\alpha_j|$, so that

$$\langle \psi \,|\, H_e(q) \,|\, \psi \rangle \approx |\alpha_1|^2 E_1(q) + |\alpha_2|^2 E_2(q)$$

and hence the classical motion in (5.1) is approximately determined by

$$\dot q = p\,, \quad \dot p = -\nabla_q\big(|\alpha_1|^2 E_1(q) + |\alpha_2|^2 E_2(q)\big)\,, \tag{5.3}$$

with a potential that is a convex linear combination of the potentials in (5.2). Unless the potentials E_j happen to be quadratic, not even the average position $\alpha_1 q_1 + \alpha_2 q_2$ is described correctly by the equations for q. The equations (5.1) are asymptotically correct, however, if we start from a pure eigenstate (where $\alpha_1 = 1$, $\alpha_2 = 0$).

This example illustrates that even very plausible-looking models must be considered with care and assessed critically by analysis and (numerical and physical) experiments.

For an asymptotic analysis of the above mixed quantum-classical model we refer to Bornemann & Schütte (1999). The quantum-mechanical part can be further restricted, assuming for example $\psi(y,t)$ in the form of a Slater determinant, thus combining classical motion and the time-dependent Hartree-Fock method. Global existence of solutions for such a model has been studied by Cancès & Le Bris (1999).

II.5.2 Quantum Dressed Classical Mechanics

Even if the approximation by a Gaussian wave packet is too rough, it can nevertheless be reused in a correction scheme, which is once more based on the time-dependent variational principle. We briefly describe such an approach due to Billing (2003). Let $q(t)$, $p(t)$ be defined by Gaussian wave packet dynamics with a diagonal width matrix with entries $c_n(t)$, possibly further simplified by using the classical equations of motion for q and p and a similar simplification in the differential equations for the widths, replacing averages by point evaluations. We search for an approximation to the wave function of the form

$$\psi(x_1,\ldots,x_N,t) \approx \sum_J a_J(t)\,\phi^{(1)}_{j_1}(x_1,t)\cdot\ldots\cdot\phi^{(N)}_{j_N}(x_N,t)\,,$$

where the sum is over a set of multi-indices $J = (j_1,\ldots,j_N)$ and the functions $\phi^{(n)}_j$ are shifted and scaled Gauss-Hermite basis functions defined by (we assume all x_n one-dimensional for simplicity)

$$\begin{aligned}\phi^{(n)}_j(x_n,t) \;&=\; \exp\Big(\frac{i}{\varepsilon}\Big(c_n(t)\,(x_n-q_n(t))^2 + p_n(t)\,(x_n-q_n(t))\Big)\Big)\cdot\\ &\qquad H_j\Big(\sqrt{\frac{2\,\mathrm{Im}\,c_n(t)}{\varepsilon}}\,(x_n-q_n(t))\Big)\end{aligned}$$

with Hermite polynomials H_j and the known Gaussian parameters $q_n(t)$, $p_n(t)$, and $c_n(t)$. The unknown coefficients $a_J(t)$ are determined by differential equations obtained from the variational principle on the *time-dependent* approximation manifold (here actually a linear space)

$$\mathcal{M}_t = \{u : u(x_1,\dots,x_N) = \sum_J a_J\, \phi^{(1)}_{j_1}(x_1,t)\cdot\ldots\cdot\phi^{(N)}_{j_N}(x_N,t),\ a_J \in \mathbb{C}\},$$

at every instant t as previously in (1.2), except that now du/dt is not sought for in the tangent space of $\mathcal{M}_t$, but as the derivative of a path $u(t) \in \mathcal{M}_t$.

This approach leads to a method which adapts the location and width of the Hermite basis functions to Gaussian wave packets that follow classical trajectories. We will consider in more detail a somewhat related, but computationally favourable approach in Chap. V.

II.5.3 Swarms of Gaussians

In a conceptually similar approach, frozen Gaussians $\gamma_k(x,t)$ first evolve independently according to the classical equations of motion for position and momentum and with the phase given by the action integral $\int_0^t \left(\frac{1}{2}|p_k|^2 - \langle V\rangle_{\gamma_k}\right) ds$, as proposed by Heller (1981). This approximation is then improved upon by taking a linear combination

$$\psi(x,t) \approx \sum_k a_k(t)\, \gamma_k(x,t)\,,$$

where the coefficients $a_k(t)$ are determined by the time-dependent variational principle:

$$\Big\langle \sum_j b_j\gamma_j \,\Big|\, \sum_k (\dot a_k\gamma_k + a_k\dot\gamma_k) - \frac{1}{i\varepsilon} H \sum_k a_k\gamma_k \Big\rangle = 0 \qquad \forall\, b=(b_j)\,.$$

This yields a linear system of differential equations for $a = (a_k)$,

$$M\dot a = \frac{1}{i\varepsilon} Ka - La$$

with the matrices $M = \big(\langle\gamma_j\,|\,\gamma_k\rangle\big)$, $L = \big(\langle\gamma_j\,|\,\dot\gamma_k\rangle\big)$, $K = \big(\langle\gamma_j\,|\,H\,|\,\gamma_k\rangle\big)$. While the L^2 norm of the approximation is conserved, the total energy and symplecticity are *not* conserved by applying the variational principle on a time-dependent approximation space as is done here, in contrast to the case of a time-independent approximation manifold as studied in Sect. II.1.3.

The above approach was mentioned by Heller (1981) and has been carried further by Ben-Nun & Martinez (1998, 2000) together with criteria when to create, or "spawn" new basis functions. It is related in spirit to particle methods in fluid dynamics; see, e.g., Monaghan (1992) and Yserentant (1997).

II.6 Quasi-Optimality of Variational Approximations

In this theoretical section we consider variational approximation on a manifold $\mathcal{M}$ and study the following question: In case the true wave function remains close to the manifold, does the time-dependent variational principle then provide a good approximation? Stated differently: Can the error of the variational approximation be bounded in terms of the error of the best approximation to the wave function on $\mathcal{M}$?

This is a familiar question in other areas of numerical analysis; cf. Céa's lemma on the optimality of Galerkin approximations of elliptic boundary value problems as stated, e.g., in Ciarlet (1991), p. 113. A positive answer to this question separates the problems of approximability of the wave function on the chosen manifold, which often is a modeling hypothesis, and the quality of the time-dependent variational principle for obtaining approximate wave functions.

Following Lubich (2005), we give a conditionally positive answer under assumptions that include, for example, the time-dependent Hartree method and its multi-configuration versions.

Assumptions. We consider the Schrödinger equation (1.1) on a Hilbert space $\mathcal{H}$, with $\hbar = 1$ in the following, and the variational approximation given by the Dirac-Frenkel principle (1.2) on the manifold $\mathcal{M}$. The Hamiltonian H is split as

$$H = A + B \tag{6.1}$$

with self-adjoint linear operators A and B where A is such that $u \in \mathcal{M}$ implies $e^{-itA}u \in \mathcal{M}$ for all t. This is satisfied if and only if A is tangential, that is,

$$Au \in \mathcal{T}_u\mathcal{M} \qquad \text{for all} \qquad u \in \mathcal{M} \cap D(A). \tag{6.2}$$

We assume that the (non-tangential) operator B is bounded:

$$\|B\varphi\| \leq \beta\,\|\varphi\| \tag{6.3}$$

for all $\varphi \in \mathcal{H}$. About the approximation manifold $\mathcal{M}$ we assume the condition (1.3) of complex linear tangent spaces $\mathcal{T}_u\mathcal{M}$, and a condition that is satisfied if $\mathcal{M}$ contains rays (cf. Theorem 1.4):

$$u \in \mathcal{T}_u\mathcal{M} \qquad \text{for all} \qquad u \in \mathcal{M}\,, \tag{6.4}$$

A bound of the curvature of $\mathcal{M}$ is formulated in terms of the orthogonal projectors $P(u) : \mathcal{H} \to \mathcal{T}_u\mathcal{M}$ and $P^\perp(u) = I - P(u)$:

$$\begin{aligned} \|\,(P(u) - P(v))\varphi\,\| &\leq \kappa\,\|u - v\| \cdot \|\varphi\| &(6.5)\\ \|\,P^\perp(v)(u - v)\,\| &\leq \kappa\,\|u - v\|^2 &(6.6) \end{aligned}$$

for all $u, v \in \mathcal{M}$ and $\varphi \in \mathcal{H}$. We assume that $P(u(t))\varphi$ is a continuously differentiable function of t in $\mathcal{H}$ for every continuously differentiable path $u(t)$ on $\mathcal{M}$ and $\varphi \in \mathcal{H}$. These assumptions will actually be needed only in a neighbourhood of the wave function or the variational approximation, so that only a local bound of the curvature rather than a global bound enters the estimates.

The initial data $\psi(0)$ is assumed to be on $\mathcal{M}$ and of unit norm. We consider a time interval on which the solution $\psi(t)$ to (1.1) remains near $\mathcal{M}$, in the sense that

$$\operatorname{dist}(\psi(t), \mathcal{M}) \le \frac{1}{2\kappa} \qquad \text{for} \quad 0 \le t \le \bar{t}\,. \tag{6.7}$$

Both the exact wave function $\psi(t)$ and the variational approximation $u(t)$ of (1.2) are required to be in the domain of H for $0 \le t \le \bar{t}$, with a bound

$$\|H\psi(t)\| \le \mu\,, \qquad \|Hu(t)\| \le \mu \quad \text{and} \quad \|Au(t)\| \le \mu\,. \tag{6.8}$$

Further we consider the distance bound $\delta \le \mu$ given by

$$\operatorname{dist}\left(H\psi(t), \mathcal{T}_{v(t)}\mathcal{M}\right) \le \delta\,, \qquad \operatorname{dist}\left(Hu(t), \mathcal{T}_{u(t)}\mathcal{M}\right) \le \delta\,, \tag{6.9}$$

where $v(t) \in \mathcal{M}$ is the nearest point to $\psi(t)$ on $\mathcal{M}$:

$$\|v(t) - \psi(t)\| = \operatorname{dist}\left(\psi(t), \mathcal{M}\right).$$

Discussion of the Assumptions. In all the examples of this chapter, A might be chosen as the kinetic energy operator T, though this might not always be the optimal choice. A more critical assumption is the boundedness of the operator B that maps outside the tangent space. It is a reasonable assumption in the Schrödinger equation of the nuclei and its Hartree and Gaussian wave packet approximations (and their multi-configuration versions). The condition is not satisfied, however, in the time-dependent Hartree-Fock method for the electronic Schrödinger equation where the Coulomb potentials are unbounded. We refer to Lubich (2005) for a corresponding result in the Coulomb case.

We have assumed the splitting (6.1) independent of time for ease of presentation, though the result would extend directly to the situation of a time-dependent splitting $H = A(t) + B(t)$. For example, in the (multi-configuration) Hartree method we might choose $A(t) = T + V_1 + \ldots + V_N$ with the mean-field potentials V_n, so that $B(t)$ becomes the difference between the given potential and the sum of the mean-field potentials. This can be expected to give more favourable error bounds than a time-independent splitting into kinetic energy and potential. For Gaussian wave packets we can split into $A(t) = T + U_{q(t)}$ with the local quadratic approximation $U_{q(t)}$ to the potential at the position $q(t)$, and the non-quadratic remainder $B(t)$, as we did in the proof of Theorem 4.4.

Condition (6.4) is satisfied for all the examples in this chapter. Conditions (6.5) and (6.6) encode curvature information of $\mathcal{M}$ in a form that is suitable for our analysis. Condition (6.7) ensures that $\psi(t)$ has a unique nearest point on $\mathcal{M}$. The regularity assumption (6.8) for $\psi(t)$ is satisfied if the initial value has such regularity. The regularity (6.8) of the approximate solution $u(t)$ needs to be ascertained, but is known to hold, e.g., for the (multi-configuration) time-dependent Hartree method when the Schrödinger equation for the nuclei has a smooth bounded potential.

The following result bounds the error of the variational approximation in terms of the best-approximation error.

Theorem 6.1 (Quasi-Optimality of Variational Approximations). *Under conditions (6.1)–(6.9), the error of the variational approximation is bounded by*

$$\|u(t)-\psi(t)\| \le d(t) + Ce^{\gamma t}\int_0^t d(s)\,ds \quad \textit{with} \quad d(t)=\operatorname{dist}(\psi(t),\mathcal{M}) \tag{6.10}$$

and with $\gamma = 2\kappa\delta$ *and* $C=\beta+3\kappa\mu$, *for* $0\le t\le \bar t$.

Though the bound (6.10) can be pessimistic in a concrete situation, it does identify sources that can make the variational approximation deviate far from optimality even if the best-approximation error $d(t)$ is small: large curvature of the approximation manifold (κ), a large effective non-separable potential in the Hamiltonian (β,δ), lack of regularity in the exact or approximate solution (μ,δ), and long time intervals (t).

Proof. The proof compares the differential equation for $u(t)$ with the equation satisfied by the best approximation $v(t)\in\mathcal{M}$ with $\|v(t)-\psi(t)\|=d(t)$.

(a) The function $v(t)$ is implicitly characterized by the condition (omitting the obvious argument t in the following)

$$P(v)\,(v-\psi)=0\,. \tag{6.11}$$

Under condition (6.7), the implicit function theorem can be used to show that this equation has a unique solution in the ball of radius $1/(2\kappa)$ around ψ, which depends continuously differentiable on t. We derive a differential equation for $v(t)$ by differentiating (6.11) with respect to t ($\dot{} = d/dt$):

$$0=P(v)(\dot v-\dot\psi)+\big(P'(v)\cdot(v-\psi)\big)\dot v \tag{6.12}$$

with $(P'(v)\cdot\varphi)\dot v=(d/dt)P(v(t))\varphi$ for $\varphi\in\mathcal{H}$. Since $\dot v\in T_v\mathcal{M}$, we have $P(v)\dot v=\dot v$, and the equation becomes

$$\Big(I+P'(v)\cdot(v-\psi)\Big)\dot v=P(v)\dot\psi\,. \tag{6.13}$$

By (6.5) and (6.7) we have

$$\|P'(v)\cdot(v-\psi)\|\le\kappa\,\|v-\psi\|\le\frac12\,,$$

so that the operator in (6.13) is invertible and

$$\dot v=P(v)\dot\psi+r(v,\psi) \qquad \text{with} \qquad \|r(v,\psi)\|\le 2\kappa\mu\,\|v-\psi\|\,. \tag{6.14}$$

Here we have used the bound (6.8), $\|\dot\psi\|=\|H\psi\|\le\mu$. Inserting (1.1) in (6.14), the equation can be written as

$$\dot v=P(v)\frac{1}{i}Hv-P(v)\frac{1}{i}H(v-\psi)+r(v,\psi)\,. \tag{6.15}$$

We will compare this differential equation with Equation (1.5) for $u(t)$, viz.,

$$\dot{u} = P(u)\frac{1}{i}Hu\,. \tag{6.16}$$

In (6.15) and in the following we tacitly assume $v(t) \in D(H) = D(A)$. (If v does not have this regularity, then the proof would proceed by replacing v by a regularized family (v_ε) with $v_\varepsilon(t) \in \mathcal{M} \cap D(H)$ and $v_\varepsilon \to v$ in $C^1([0,\bar{t}],\mathcal{H})$ as $\varepsilon \to 0$. Applying the arguments below to v_ε and letting $\varepsilon \to 0$ in the final estimate then gives the result.)

(b) We form the difference of (6.16) and (6.15), take the inner product with $u - v$ and consider the real part. We then have

$$\|u-v\| \cdot \frac{d}{dt}\|u-v\| = \frac{1}{2}\frac{d}{dt}\|u-v\|^2 = \mathrm{Re}\,\langle u - v \,|\, \dot{u} - \dot{v}\rangle = I + II + III$$

with

$$\begin{aligned} I &= -\mathrm{Re}\,\langle u - v \,|\, P(u)iHu - P(v)iHv\rangle \\ II &= -\mathrm{Re}\,\langle u - v \,|\, P(v)iH(v-\psi)\rangle \\ III &= -\mathrm{Re}\,\langle u - v \,|\, r(v,\psi)\rangle\,. \end{aligned}$$

(c) Using the self-adjointness of $H = A + B$ and condition (6.2), which implies $P^\perp(v)iAv = 0$, we write

$$\begin{aligned} I &= \mathrm{Re}\,\langle u - v \,|\, P^\perp(u)iHu - P^\perp(v)iHv\rangle \\ &= \mathrm{Re}\,\langle u - v \,|\, P^\perp(u)iHu\rangle - \mathrm{Re}\,\langle u - v \,|\, P^\perp(v)iBv\rangle\,. \end{aligned}$$

To treat the expression II, we split

$$II = -\mathrm{Re}\,\langle u - v \,|\, P(v)iA(v-\psi)\rangle - \mathrm{Re}\,\langle u - v \,|\, P(v)iB(v-\psi)\rangle\,.$$

It is in the first term that condition (6.4) is used. This condition implies $P(v)v = v$ and hence, by (6.11),

$$v = P(v)\psi, \qquad v - \psi = P^\perp(v)(v-\psi) = -P^\perp(v)\psi\,.$$

It follows that

$$\langle v \,|\, P(v)iA(v-\psi)\rangle = -\langle v \,|\, P(v)iAP^\perp(v)\psi\rangle = \langle P^\perp(v)iAv \,|\, \psi\rangle\,.$$

Since $P^\perp(v)iAv = 0$ by (6.2), we obtain

$$\langle v \,|\, P(v)iA(v-\psi)\rangle = 0\,. \tag{6.17}$$

Similarly, (6.2) implies

$$\langle u \,|\, iAP^\perp(u)(v-\psi)\rangle = 0\,. \tag{6.18}$$

Now, (6.17) yields

$$\langle u - v \,|\, P(v)iA(v-\psi)\rangle = \langle u \,|\, iA(v-\psi)\rangle - \langle u - v \,|\, P^\perp(v)iA(v-\psi)\rangle\,.$$

With (6.18) and once again $P^\perp(v)iAv = 0$, we obtain

$$\begin{aligned}&\langle u - v \,|\, P(v)iA(v-\psi)\rangle \\ &= -\langle u \,|\, iA(P^\perp(u) - P^\perp(v))(v-\psi)\rangle + \langle u - v \,|\, P^\perp(v)iA\psi\rangle\,.\end{aligned}$$

Using that A and $P^\perp(v) = P^\perp(v)^2$ are self-adjoint and that $A = H - B$ by (6.1), this is rewritten as

$$\begin{aligned}&\langle u - v \,|\, P(v)iA(v-\psi)\rangle \\ &= -\langle iAu \,|\, (P(u) - P(v))(v-\psi)\rangle + \langle P^\perp(v)(u-v) \,|\, P^\perp(v)iH\psi\rangle \\ &\quad -\langle u - v \,|\, P^\perp(v)iB\psi\rangle\,.\end{aligned}$$

We then arrive at the basic equation of the proof,

$$\begin{aligned} I + II \;=\;& \mathrm{Re}\,\langle P^\perp(u)(u-v) \,|\, P^\perp(u)iHu\rangle \\ &- \mathrm{Re}\,\langle u - v \,|\, iB(v-\psi)\rangle \\ &+ \mathrm{Re}\,\langle iAu \,|\, (P(u) - P(v))(v-\psi)\rangle \\ &- \mathrm{Re}\,\langle P^\perp(v)(u-v) \,|\, P^\perp(v)iH\psi\rangle\,.\end{aligned}$$

With (6.3)–(6.9) we thus obtain

$$\begin{aligned} |I + II| \;\le\;& \kappa\,\|u-v\|^2\cdot\delta + \|u-v\|\cdot\beta\,\|v-\psi\| \\ &+ \mu\cdot\kappa\,\|u-v\|\cdot\|v-\psi\| + \kappa\,\|u-v\|^2\cdot\delta \\ =\;& 2\kappa\delta\,\|u-v\|^2 + (\beta+\kappa\mu)\,\|u-v\|\cdot\|v-\psi\|\,.\end{aligned}$$

(d) Together with (6.14) for bounding III, this estimate gives

$$\frac{d}{dt}\,\|u-v\| \le \gamma\|u-v\| + C\|v-\psi\|$$

with $\gamma = 2\kappa\delta$ and $C = \beta + 3\kappa\mu$. The Gronwall inequality then implies

$$\|u(t) - v(t)\| \le Ce^{\gamma t}\int_0^t \|v(s) - \psi(s)\|\,ds\,, \tag{6.19}$$

and the triangle inequality for $u - \psi = (u-v) + (v-\psi)$ together with $d = \|v-\psi\|$ yield the result. □

Chapter III.
Numerical Methods for the Time-Dependent Schrödinger Equation

This chapter deals with numerical methods for linear time-dependent Schrödinger equations, of low to moderate dimension (less than 10, say). Although the emphasis is on time-dependent aspects, we begin with a section on space discretization, where we describe the Galerkin and collocation approaches on the important examples of Hermite and Fourier bases, including their extension to higher dimensions using hyperbolic cross approximations and sparse grids for which the computational work grows only mildly with the dimension.

We then turn to time-stepping methods: polynomial approximations to the exponential of the Hamiltonian based on the Lanczos method or on Chebyshev polynomials, and splitting methods and their high-order refinements by composition. We conclude the chapter with a brief look at integrators for Schrödinger equations with a time-varying potential.

The time-dependent Schrödinger equation considered in this chapter is in $d \geq 1$ space dimensions, has $\hbar = 1$ and reads

$$i\,\frac{\partial \psi}{\partial t} = H\psi\,, \qquad H = T + V\,, \tag{0.1}$$

with the kinetic energy operator $T = -\frac{1}{2\mu}\Delta$ for a positive mass parameter μ and with a potential $V(x)$.

III.1 Space Discretization by Spectral Methods

We follow two tracks (among many possible) for the discretization of the Schrödinger equation in space: the Galerkin method with a basis of Hermite functions and collocation with trigonometric polynomials. Both cases are instances of spectral or pseudospectral methods, which are of common use in many application areas; see, e.g., Canuto, Hussaini, Quarteroni & Zang (2006), Fornberg (1996), Gottlieb & Orszag (1977), and Trefethen (2000). Both cases are studied here for the Schrödinger equation in one and several dimensions, with the extension to higher dimensions by hyperbolically reduced tensor product bases.

III.1.1 Galerkin Method, 1D Hermite Basis

Galerkin Method. We consider an approximation space $\mathcal{V}_K \subset L^2(\mathbb{R}^d)$ spanned by K basis functions $\varphi_0, \ldots, \varphi_{K-1}$. We determine an approximate wave function $\psi_K(t) \in \mathcal{V}_K$ by the condition that at every instant t, its time derivative is determined by the condition

$$\frac{d\psi_K}{dt} \in \mathcal{V}_K \quad \text{such that} \quad \Big\langle \varphi \,\Big|\, i\,\frac{d\psi_K}{dt} - H\psi_K \Big\rangle = 0 \qquad \forall \varphi \in \mathcal{V}_K\,. \tag{1.1}$$

This is, of course, the time-dependent variational principle (II.1.2) on the linear approximation space $\mathcal{V}_K$. In particular, we know from Sect. II.1 that norm, energy and symplectic structure are preserved. Writing the approximation as a linear combination of basis functions

$$\psi_K(t) = \sum_{k=0}^{K-1} c_k(t)\varphi_k \tag{1.2}$$

and inserting in (1.1), we obtain for the time-dependent coefficient vector $c = (c_k)$ the linear system of ordinary differential equations

$$i\,M_K\dot{c} = H_K c \tag{1.3}$$

with the matrices

$$M_K = \big(\langle \varphi_j \,|\, \varphi_k\rangle\big)_{j,k=0}^{K-1}, \qquad H_K = \big(\langle \varphi_j \,|\, H \,|\, \varphi_k\rangle\big)_{j,k=0}^{K-1}. \tag{1.4}$$

The matrix M_K becomes the identity matrix in the case of an orthonormal basis, where $\langle \varphi_j \,|\, \varphi_k\rangle = \delta_{jk}$.

Hermite Basis in 1D. After a suitable rescaling and shift $x \to \alpha x + \beta$, this is the choice of basis functions

$$\varphi_k(x) = \frac{1}{\pi^{1/4}}\,\frac{1}{\sqrt{2^k k!}}\,H_k(x)\,e^{-x^2/2}\,. \tag{1.5}$$

Here, $H_k(x)$ is the Hermite polynomial of degree k, which is the kth orthogonal polynomial with respect to the weight function e^{-x^2} on $\mathbb{R}$; see, e.g., Abramowitz & Stegun (1965). While formula (1.5) does not fail to impress, it is neither useful for computations nor for understanding the approximation properties of this basis. We therefore now turn to another way of writing the Hermite functions φ_k, which also provides some motivation for the use of this basis.

Ladder Operators. We recall the canonical commutator relation (I.4.8) between the one-dimensional position operator q given by $(q\psi)(x) = x\psi(x)$ and the momentum operator $p = -i\,d/dx$:

$$\frac{1}{i}\,[q,p] = 1.$$

It follows that Dirac's *ladder operators* defined by

$$A = \frac{1}{\sqrt{2}}\,(q + ip)\,, \qquad A^\dagger = \frac{1}{\sqrt{2}}\,(q - ip) \tag{1.6}$$

satisfy the relations

$$A^\dagger A = \frac{1}{2}\,(p^2+q^2) - \frac{1}{2}\,, \qquad AA^\dagger = \frac{1}{2}\,(p^2+q^2) + \frac{1}{2}\,, \tag{1.7}$$

so that $A^\dagger A$ and $AA^\dagger$ have the same eigenfunctions as the Hamiltonian of the harmonic oscillator, $\frac{1}{2}\,(p^2+q^2)$. We also note

$$AA^\dagger = A^\dagger A + 1\,. \tag{1.8}$$

Moreover, $A^\dagger$ is adjoint to A on the Schwartz space $\mathcal{S}$ of smooth rapidly decaying functions:

$$\langle A^\dagger \varphi \,|\, \psi\rangle = \langle \varphi \,|\, A\psi\rangle \qquad \forall\, \varphi, \psi \in \mathcal{S}\,. \tag{1.9}$$

Harmonic Oscillator Eigenfunctions via the Dirac Ladder. We note that the Gaussian $\phi_0(x) = e^{-x^2/2}$ is in the kernel of A: $A\phi_0 = 0$. Moreover, it is checked that multiples of ϕ_0 are the only L^2 functions in the kernel of A, whereas $A^\dagger$ has only the trivial kernel 0. With (1.8) it follows that

$$AA^\dagger \phi_0 = A^\dagger A \phi_0 + \phi_0 = \phi_0\,,$$

and hence ϕ_0 is an eigenfunction of $AA^\dagger$ to the eigenvalue 1. Applying the operator $A^\dagger$ to both sides of this equation, we see that $\phi_1 = A^\dagger\phi_0$ is an eigenfunction of $A^\dagger A$ to the eigenvalue 1, and again by (1.8) an eigenfunction of $AA^\dagger$ to the eigenvalue 2. We continue in this way to construct successively $\phi_{k+1} = A^\dagger \phi_k$ for $k \ge 0$. We thus obtain eigenfunctions ϕ_k to $A^\dagger A$, with eigenvalue k, and to $AA^\dagger$, with eigenvalue $k+1$. These eigenfunctions are not yet normalized. To achieve this, we note that by (1.9),

$$\|A^\dagger \phi_k\|^2 = \langle A^\dagger\phi_k \,|\, A^\dagger\phi_k\rangle = \langle \phi_k \,|\, AA^\dagger\phi_k\rangle = (k+1)\,\|\phi_k\|^2\,.$$

We therefore obtain eigenfunctions to $AA^\dagger$ and $A^\dagger A$ of unit L^2 norm by setting

$$\varphi_0(x) = \frac{1}{\pi^{1/4}}\, e^{-x^2/2} \tag{1.10}$$

and

$$\varphi_{k+1} = \frac{1}{\sqrt{k+1}}\, A^\dagger \varphi_k \qquad \text{for} \quad k \ge 0\,. \tag{1.11}$$

Since $A\varphi_{k+1} = \frac{1}{\sqrt{k+1}}\, AA^\dagger\varphi_k = \sqrt{k+1}\,\varphi_k$, we also have (replacing $k+1$ by k)

$$\varphi_{k-1} = \frac{1}{\sqrt{k}}\, A\varphi_k \qquad \text{for} \quad k \ge 0\,. \tag{1.12}$$

These relations explain the names of *raising operator* and *lowering operator* for $A^\dagger$ and A, respectively, and of *ladder operators* for both of them. Multiplying (1.11) by $\sqrt{k+1}$ and (1.12) by $\sqrt{k}$, summing the resulting formulas and using the definitions of A and $A^\dagger$, we obtain the three-term recurrence relation

$$\sqrt{k+1}\,\varphi_{k+1}(x) = \sqrt{2}\,x\,\varphi_k(x) - \sqrt{k}\,\varphi_{k-1}(x) \qquad \text{for} \quad k \ge 0\,, \tag{1.13}$$

with $\varphi_{-1}(x) = 0$. This allows us to evaluate $\varphi_k(x)$ at any required point x. We state essential properties of these functions.

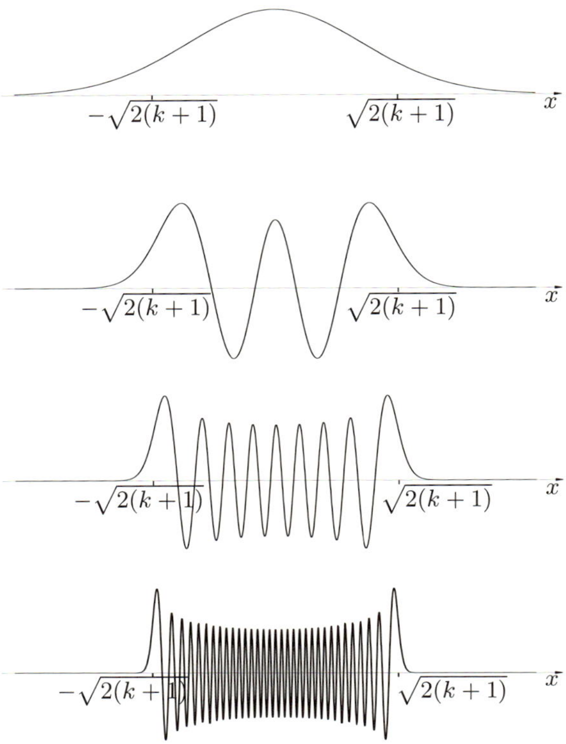

Fig. 1.1. Hermite functions φ_k for $k = 0, 4, 16, 64$.

Theorem 1.1 (Hermite Functions). *The functions φ_k defined by (1.10) and (1.11) form a complete L^2-orthonormal set of functions, the eigenfunctions of the harmonic oscillator Hamiltonian $\frac{1}{2}(p^2 + q^2)$. They are identical to the Hermite functions given by (1.5).*

Proof. From the above construction it is clear that each φ_k is an oscillator eigenfunction to the eigenvalue $k + \frac{1}{2}$. As normalized eigenfunctions of a self-adjoint operator, the φ_k are orthonormal. It is also clear from the recurrence relation that φ_k is a polynomial of degree k times $e^{-x^2/2}$. By the orthonormality, this polynomial must be a multiple of the kth Hermite polynomial, which yields (1.5). For the proof of completeness we refer to Thaller (2000), Sect. 7.8. □

The completeness together with orthonormality yields that every function $f \in L^2(\mathbb{R})$ can be expanded as the series

$$f = \sum_{k=0}^{\infty} \langle \varphi_k \mid f \rangle \, \varphi_k \,, \tag{1.14}$$

where the convergence of the series is understood as convergence of the partial sums in the L^2 norm.

Approximation Properties. We denote by P_K the orthogonal projector onto $\mathcal{V}_K = \mathrm{span}(\varphi_0, \dots, \varphi_{K-1})$, given by

$$P_K f = \sum_{k<K} \langle \varphi_k \,|\, f \rangle \, \varphi_k .$$

This is the best approximation to f in $\mathcal{V}_K$ with respect to the L^2 norm. We have the following approximation result, for which we recall $A = \frac{1}{\sqrt{2}}(x + d/dx)$.

Theorem 1.2 (Approximation by Hermite Functions). *For every integer $s \le K$ and every function f in the Schwartz space $\mathcal{S}$,*

$$\|f - P_K f\| \le \frac{1}{\sqrt{K(K-1)\dots(K-s+1)}} \, \|A^s f\| .$$

Given sufficient smoothness and decay of the function, the approximation error thus decays as $\mathcal{O}(K^{-s/2})$ for growing K and any fixed s.

Proof. Using subsequently (1.14), (1.11) and (1.9) we obtain

$$\begin{aligned} f - P_K f &= \sum_{k \ge K} \langle \varphi_k \,|\, f \rangle \, \varphi_k \\ &= \sum_{k \ge K} \frac{1}{\sqrt{k(k-1)\dots(k-s+1)}} \langle (A^\dagger)^s \varphi_{k-s} \,|\, f \rangle \, \varphi_k \\ &= \sum_{k \ge K} \frac{1}{\sqrt{k(k-1)\dots(k-s+1)}} \langle \varphi_{k-s} \,|\, A^s f \rangle \, \varphi_k . \end{aligned}$$

By orthonormality, this yields

$$\begin{aligned} \|f - P_K f\|^2 &\le \frac{1}{K(K-1)\dots(K-s+1)} \sum_{j \ge 0} \big| \langle \varphi_j \,|\, A^s f \rangle \big|^2 \\ &= \frac{1}{K(K-1)\dots(K-s+1)} \, \|A^s f\|^2 , \end{aligned}$$

which is the desired result. □

Since the set of linear combinations of shifted Gaussians is known to be dense in $L^2(\mathbb{R})$ (e.g., Thaller, 2000, p. 40), it is instructive to see the action of A^s on $e^{-(x-a)^2/2}$. A short calculation yields $A\, e^{-(x-a)^2/2} = \frac{1}{\sqrt{2}}\, a\, e^{-(x-a)^2/2}$ and hence

$$A^s \, e^{-(x-a)^2/2} = 2^{-s/2} \, a^s \, e^{-(x-a)^2/2} .$$

No surprise, the approximation of $e^{-(x-a)^2/2}$ by Hermite functions φ_k centered at 0 is slow to converge for large shifts $|a| \gg 1$. According to Theorem 1.2, the error becomes small from $K > \frac{e}{2} a^2$ onwards (on choosing $s = K$ and using Stirling's formula for $K!$).

Error of the Galerkin Method with Hermite Basis in 1D. We are now in the position to prove the following error bound. For a related result we refer to Faou & Gradinaru (2007).

Theorem 1.3 (Galerkin Error). *Consider the Galerkin method with the one-dimensional Hermite basis* $(\varphi_0,\dots,\varphi_{K-1})$*, applied to a 1D Schrödinger equation (0.1) with a potential* $V(x)=(1+x^2)B(x)$ *with bounded* B*, with initial value* $\psi_K(0)=P_K\psi(0)$*. Then, if the exact solution is in* $D(A^{s+2})$ *for some integer* $s\le K/2$*, the error is bounded by*

$$\|\psi_K(t)-\psi(t)\|\le C\,K^{-s/2}\,(1+t)\max_{0\le\tau\le t}\|A^{s+2}\psi(\tau)\|\,,$$

where C *is independent of* K *and* t*, is bounded by* $C\le c\,2^{s/2}$ *in dependence of* s*, and depends linearly on the bound of* B*.*

Proof. (a) We write the Galerkin equation (1.1) as

$$i\dot\psi_K=P_KHP_K\psi_K$$

with the Hermitian matrix P_KHP_K, and the Schrödinger equation (0.1), acted on by P_K, as

$$iP_K\dot\psi=P_KHP_K\,P_K\psi+P_KHP_K^\perp\psi\,,$$

where $P_K^\perp=I-P_K$ is the complementary orthogonal projection. Subtracting the two equations and taking the inner product with $\psi_K-P_K\psi$ yields, by the same argument as in the proof of Theorem II.1.5,

$$\|\psi_K(t)-P_K\psi(t)\|\le\|\psi_K(0)-P_K\psi(0)\|+\int_0^t\|P_KHP_K^\perp\psi(\tau)\|\,d\tau\,.$$

We show in part (b) of the proof that

$$\|P_KHP_K^\perp\psi\|\le C\,K^{-s/2}\,\|A^{s+2}\psi\|\,. \tag{1.15}$$

The result then follows together with Theorem 1.2, applied with $s+2$ instead of s, to estimate $\psi(t)-P_K\psi(t)$.

(b) It remains to prove (1.15). We recall that $H=\frac{1}{2\mu}p^2+B(1+q^2)$. By (1.6) we have

$$p^2=-\frac12(A-A^\dagger)^2\,,\qquad q^2=\frac12(A+A^\dagger)^2\,.$$

With (1.11) and (1.12) this gives

$$\begin{aligned}p^2\varphi_k&=-\frac12\big(\sqrt{k(k-1)}\,\varphi_{k-2}-(2k+1)\varphi_k+\sqrt{(k+2)(k+1)}\,\varphi_{k+2}\big)\\ q^2\varphi_k&=\frac12\big(\sqrt{k(k-1)}\,\varphi_{k-2}+(2k+1)\varphi_k+\sqrt{(k+2)(k+1)}\,\varphi_{k+2}\big)\,.\end{aligned}$$

This yields, with $c_k=\langle\varphi_k\,|\,\psi\rangle$,

$$P_Kp^2P_K^\perp\psi=c_K\sqrt{K(K-1)}\,\varphi_{K-2}+c_{K+1}\sqrt{(K+1)K}\,\varphi_{K-1}\,.$$

Estimating the coefficients c_k as in the proof of Theorem 1.2 with $s+2$ instead of s, we obtain

$$\|P_K p^2 P_K^\perp \psi\| \le C\, K^{-s/2}\, \|A^{s+2}\psi\|\,.$$

Similarly, we get

$$\|q^2 P_K^\perp \psi\| \le C\, K^{-s/2}\, \|A^{s+2}\psi\|\,.$$

Together with the boundedness of B, these two estimates imply the bound (1.15). □

We remark that from Theorem II.1.5, we can alternatively obtain an a posteriori error bound $C\,K^{-s/2}\,t\,\max_{0\le\tau\le t}\big(\|A^{s+2}\psi_K(\tau)\|+\|A^{s+2}B\psi_K(\tau)\|\big)$, where the approximate solution ψ_K instead of the exact solution ψ appears in the estimate.

Computation of the Matrix Elements. To compute the entries of the matrix H_K of (1.4), we split into the harmonic oscillator and the remaining potential,

$$H = D + W \equiv \frac{1}{2\mu}(p^2+q^2) + \Big(V - \frac{1}{2\mu}\,q^2\Big)\,.$$

and consider the corresponding matrices

$$D_K = \big(\langle\varphi_j\,|\,D\,|\,\varphi_k\rangle\big)_{j,k=0}^{K-1}, \qquad W_K = \big(\langle\varphi_j\,|\,W\,|\,\varphi_k\rangle\big)_{j,k=0}^{K-1}.$$

By Theorem 1.1, D_K is diagonal with entries $d_k = (k+\frac{1}{2})/\mu$. To compute W_K, we use *Gauss–Hermite quadrature*, that is, Gaussian quadrature for the weight function e^{-x^2} over $\mathbb{R}$ (see, e.g., Gautschi 1997): for $M \ge K$, let x_i ($i = 1,\dots,M$) be the zeros of the Mth Hermite polynomial $H_M(x)$. With the corresponding weights w_i or $\omega_i = w_i\, e^{x_i^2}$, the quadrature formula

$$\int_{-\infty}^{\infty} e^{-x^2}\, h(x)\, dx \approx \sum_{i=1}^{M} w_i\, h(x_i) \quad\text{or}\quad \int_{-\infty}^{\infty} f(x)\, dx \approx \sum_{i=1}^{M} \omega_i\, f(x_i)$$

is exact for all polynomials h of degree up to $2M-1$. If $f(x) = g(x)\cdot e^{-x^2/2}$ with a function $g \in L^2(\mathbb{R})$ for which the coefficients $c_k = \langle\varphi_k\,|\,g\rangle$ in the Hermite expansion (1.14) of g satisfy $|c_k| \le C\,(1+k)^{-r}$ with $r > 1$, we then obtain that the quadrature error is bounded by $\mathcal{O}(M^{-r})$.

We thus approximate

$$\langle\varphi_j\,|\,W\,|\,\varphi_k\rangle \approx \sum_{i=1}^{M} \omega_i\, \varphi_j(x_i)\, W(x_i)\, \varphi_k(x_i)\,, \tag{1.16}$$

using M evaluations of the potential for all K^2 matrix elements, and evaluating $\varphi_j(x_i)$ via the recurrence relation (1.13). To obtain all matrix elements with good accuracy, one would have to choose M distinctly larger than K, but in practice a popular choice is $M = K$. Though the lower right block in the matrix is then inaccurate, this does not impair the asymptotic accuracy of the overall numerical method for large K, since the inaccurate matrix elements only meet with the small coefficients that correspond to high-order Hermite functions. This observation can be turned into rigorous estimates with the arguments of the above proofs.

III.1.2 Higher Dimensions: Hyperbolic Cross and Sparse Grids

We now turn to the Galerkin method with a tensor-product Hermite basis for the d-dimensional Schrödinger equation (0.1).

Full Tensor-Product Basis. The theory of the preceding section immediately extends to a full tensor-product basis of Hermite functions: for all multi-indices $k = (k_1, \dots, k_d)$ with integers $0 \le k_n < K$, take the product functions

$$\varphi_{(k_1,\dots,k_d)}(x_1, \dots, x_d) = \varphi_{k_1}(x_1) \dots \varphi_{k_d}(x_d)$$

or briefly

$$\varphi_k = \varphi_{k_1} \otimes \dots \otimes \varphi_{k_d} \tag{1.17}$$

as the basis functions in the Galerkin method. While this is theoretically satisfactory, it is computationally infeasible in higher dimensions: the number of basis functions, the number of coefficients, the computational work all grow like K^d, exponentially with the dimension d to the large base K.

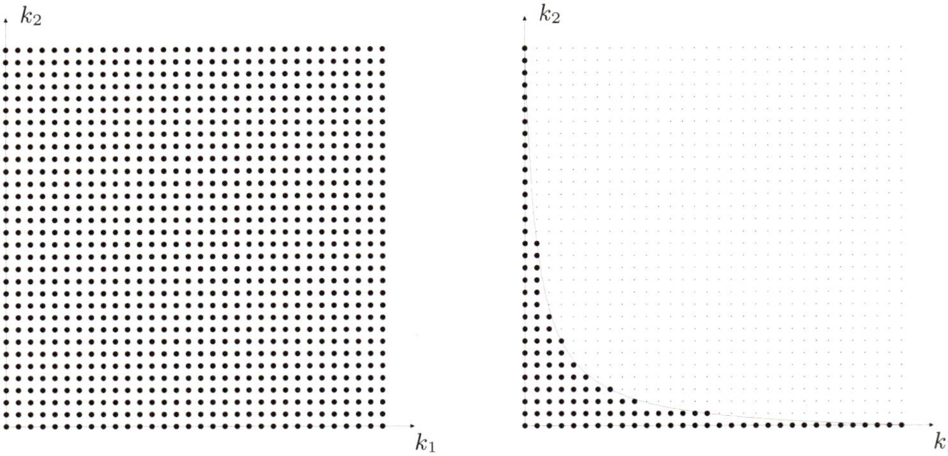

Fig. 1.2. Full and hyperbolically reduced tensor basis ($K = 32$).

Hyperbolic Reduced Tensor-Product Basis. Instead of taking *all* tensor products with $k_j < K$, we only take a subset of multi-indices: for a bound K, let the hyperbolic multi-index set $\mathcal{K}$ be given as

$$\mathcal{K} = \mathcal{K}(d, K) = \Big\{(k_1, \dots, k_d) \,:\, k_n \ge 0,\ \prod_{n=1}^{d} (1 + k_n) \le K\Big\}. \tag{1.18}$$

This is illustrated for $d = 2$ and $K = 32$ in Fig. 1.2. Taking only the tensor products φ_k of (1.17) with $k \in \mathcal{K}$ as the basis functions in the Galerkin method greatly reduces their number:

Lemma 1.4. *The number $N(d,K)$ of multi-indices in $\mathcal{K}(d,K)$ is bounded by*

$$N(d,K) \le K\,(\log K)^{d-1}\,. \tag{1.19}$$

Proof. We clearly have $N(1,K)=K$. We then note

$$N(2,K) \le \frac{K}{1} + \frac{K}{2} + \frac{K}{3} \cdots + \frac{K}{K} \le K \log K\,,$$

where the terms in the sum correspond to $k_2 = 0,1,\dots,K-1$, respectively. In general, we have

$$N(d,K) \le N(d-1,K) + N(d-1,K/2) + \dots + N(d-1,K/K)\,,$$

which by induction leads to the stated bound. □

Computations with the Galerkin method on the reduced tensor-product approximation space

$$\mathcal{V}_{\mathcal{K}} = \operatorname{span}\{\varphi_k : k \in \mathcal{K}\} \tag{1.20}$$

thus appear to become feasible up to fairly large dimension d.

Approximation Properties. Can we still get decent approximations on this reduced space? As we show next, this is possible under more stringent regularity assumptions on the functions to be approximated. We denote by $P_{\mathcal{K}}$ the orthogonal projector onto $\mathcal{V}_{\mathcal{K}}$, given by

$$P_{\mathcal{K}} f = \sum_{k\in\mathcal{K}} \langle \varphi_k \,|\, f\rangle\, \varphi_k\,.$$

We let $A_n = \frac{1}{\sqrt{2}}(x_n + d/dx_n)$ and for a multi-index $\sigma = (\sigma_1,\dots,\sigma_d)$, we denote $A^\sigma = A_1^{\sigma_1}\dots A_d^{\sigma_d}$. We then have the following approximation result.

Theorem 1.5 (Approximation by the Reduced Tensor Hermite Basis). *For every fixed integer s and every function f in the Schwartz space $\mathcal{S}(\mathbb{R}^d)$,*

$$\| f - P_{\mathcal{K}} f \| \le C(s,d)\, K^{-s/2} \max_{|\sigma|_\infty \le s} \| A^\sigma f\|\,,$$

where the maximum is taken over all $\sigma = (\sigma_1,\dots,\sigma_d)$ with $0 \le \sigma_n \le s$ for each n.

Proof. For every multi-index $k=(k_1,\dots,k_d)$ we define the multi-index $\sigma(k)$ by the condition $k_n - \sigma(k)_n = (k_n - s)_+$ (with $a_+ = \max\{a,0\}$) for all $n=1,\dots,d$, and note that $0 \le \sigma(k)_n \le s$. Similar to the proof of Theorem 1.2 we have

$$\begin{aligned} f - P_{\mathcal{K}} f &= \sum_{k\notin\mathcal{K}} \langle \varphi_k \,|\, f\rangle\, \varphi_k \\ &= \sum_{k\notin\mathcal{K}} a_{k,s}\, \langle (A^\dagger)^{\sigma(k)} \varphi_{k-\sigma(k)} \,|\, f\rangle\, \varphi_k \\ &= \sum_{k\notin\mathcal{K}} a_{k,s}\, \langle \varphi_{k-s} \,|\, A^{\sigma(k)} f\rangle\, \varphi_k\,, \end{aligned}$$

where the coefficients $a_{k,s}$ come about by (1.11) and are given as

$$a_{k,s} = \prod_{n=1}^{d} \frac{1}{\sqrt{(1+(k_n-1)_+)\dots(1+(k_n-s)_+)}}.$$

They satisfy, for $k \notin \mathcal{K}$,

$$|a_{k,s}|^2 \le \frac{c(s,d)}{K^s}, \tag{1.21}$$

because by the definition (1.18) of $\mathcal{K}$ we have the bound, for $k \notin \mathcal{K}$ and with $r = 1, \dots, s$,

$$\prod_{n=1}^{d} (1+(k_n-r)_+) \ge K \prod_{n=1}^{d} \frac{1+(k_n-r)_+}{1+k_n} \ge K\,(r+1)^{-d}.$$

By orthonormality, (1.21) yields

$$\|f - P_{\mathcal{K}} f\|^2 \le \frac{c(s,d)}{K^s} \sum_k \left|\langle \varphi_k \,|\, A^{\sigma(k)} f\rangle\right|^2.$$

Since there are s^d different possible values of $\sigma(k)$, a crude estimation yields

$$\|f - P_{\mathcal{K}} f\|^2 \le \frac{s^d\, c(s,d)}{K^s} \max_{|\sigma|_\infty \le s} \|A^\sigma f\|^2,$$

which is the stated result. □

We note that for a shifted d-dimensional Gaussian $e^{-|x-a|^2/2}$, we have the relation $A^\sigma e^{-|x-a|^2/2} = (a/\sqrt{2})^\sigma e^{-|x-a|^2/2}$, and so we now need $K \gg \prod_{n=1}^{d}(1+|a_n|^2)$ to obtain a good approximation.

Error of the Galerkin Method with Reduced Tensor Hermite Basis. With the proof of Theorem 1.3 we then obtain the following result from Theorem 1.5.

Theorem 1.6 (Galerkin Error). *Consider the Galerkin method with the hyperbolically reduced tensor Hermite basis applied to a d-dimensional Schrödinger equation (0.1) with a potential $V(x) = (1+|x|^2)B(x)$ with bounded B, with initial value $\psi_{\mathcal{K}}(0) = P_{\mathcal{K}}\psi(0)$. Then, for any fixed integer s the error is bounded by*

$$\|\psi_{\mathcal{K}}(t) - \psi(t)\| \le C(s,d)\, K^{-s/2}\,(1+t) \max_{0\le\tau\le t} \max_{|\sigma|_\infty \le s+2} \|A^\sigma \psi(\tau)\|$$

with the maximum over all $\sigma = (\sigma_1, \dots, \sigma_d)$ with $0 \le \sigma_n \le s+2$ for each n. □

Numerical Integration Using Sparse Grids. The matrix elements $\langle \varphi_j \,|\, H \,|\, \varphi_k \rangle$ for $j, k \in \mathcal{K}$ contain high-dimensional integrals. These can be approximated by numerical integration on sparse grids, following Smolyak (1963), Zenger (1991), Gerstner & Griebel

(1998) and using an adaptation that takes care of the increasingly oscillatory behaviour of the high-order Hermite functions.

We describe Smolyak's sparse grid quadrature when based on one-dimensional Gauss–Hermite quadrature in every coordinate direction. For $\ell = 0, 1, 2, \ldots$, let x_i^ℓ denote the zeros of the Hermite polynomial of degree 2^ℓ, and let w_i^ℓ be the corresponding weights and $\omega_i^\ell = w_i^\ell \, e^{(x_i^\ell)^2}$, so that we have the one-dimensional 2^ℓ-point Gauss–Hermite quadrature formula

$$Q_\ell f = \sum_{i=1}^{2^\ell} \omega_i^\ell \, f(x_i^\ell) \approx \int_{-\infty}^{\infty} f(x)\, dx \,.$$

We introduce the difference formulas between successive levels,

$$\Delta_\ell f = Q_\ell f - Q_{\ell-1} f \,,$$

and for the lowest level we set $\Delta_0 f = Q_0 f$. The full tensor quadrature approximation at level L to a d-dimensional integral $\int_{\mathbb{R}^d} f(x_1, \ldots, x_d)\, dx_1 \ldots dx_d$ reads

$$Q_L \otimes \ldots \otimes Q_L f = \sum_{i_1=1}^{2^L} \cdots \sum_{i_d=1}^{2^L} \omega_{i_1}^L \ldots \omega_{i_d}^L \, f(x_{i_1}^L, \ldots, x_{i_d}^L) \,,$$

which can be rewritten as

$$Q_L \otimes \ldots \otimes Q_L f = \sum_{\ell_1=0}^{L} \cdots \sum_{\ell_d=0}^{L} \Delta_{\ell_1} \otimes \ldots \otimes \Delta_{\ell_d} f \tag{1.22}$$

and uses $(2^L)^d$ grid points at which f is evaluated. This number is substantially reduced in *Smolyak's algorithm*, which neglects all contributions from the difference terms with $\ell_1 + \ldots + \ell_d > L$ and thus arrives at the quadrature formula

$$\sum_{\ell_1+\ldots+\ell_d \le L} \Delta_{\ell_1} \otimes \ldots \otimes \Delta_{\ell_d} f \approx \int_{\mathbb{R}^d} f(x_1, \ldots, x_d)\, dx_1 \ldots dx_d \,. \tag{1.23}$$

Here, f is evaluated only at the points of the *sparse grid*

$$\Gamma_L^d = \{ (x_{i_1}^{\ell_1}, \ldots, x_{i_d}^{\ell_d}) : \ell_1 + \ldots + \ell_d \le L \} \,,$$

which has only $\mathcal{O}(2^L \cdot L^{d-1})$ points; as an illustration see Fig. 1.3 for $L = 5$ and $d = 2$. If $f(x) = g(x) \cdot e^{-|x|^2/2}$ with a function $g \in L^2(\mathbb{R}^d)$ for which the coefficients $c_m = \langle \varphi_m \,|\, g \rangle$ in the Hermite expansion of g satisfy

$$|c_m| \le C \prod_{n=1}^{d} (1 + m_n)^{-r} \tag{1.24}$$

with $r > 1$, then the contribution of the omitted terms with $\ell_1 + \ldots + \ell_d > L$ and hence the quadrature error can be shown, by a tedious exercise, to be bounded by $\mathcal{O}((2^L)^{-r})$.

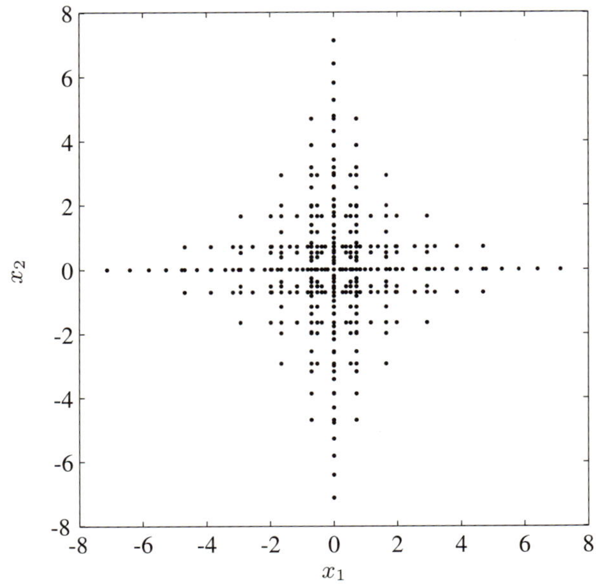

Fig. 1.3. Gauss–Hermite sparse grid ($L = 5$, $d = 2$).

Remark. A disadvantage of Gauss–Hermite quadrature formulas is the fact that they are not nested: the quadrature points of level $\ell - 1$ are not a subset of those of level ℓ. As an alternative, which will not be explored here, one might consider transformation to a finite interval and using the trapezoidal rule or Clenshaw-Curtis quadrature there. With a nested quadrature, the sparse grid contains approximately half as many grid points as for the case of a non-nested basic quadrature formula with the same number of quadrature points. It is not clear if the otherwise excellent properties of Gauss–Hermite quadrature are indeed offset by nested quadratures for suitably truncated or transformed integrals.

Computation of the Matrix Elements. The integrand f_{jk} in the matrix element

$$\langle \varphi_j \,|\, W \,|\, \varphi_k \rangle = \int_{\mathbb{R}^d} \varphi_j(x)\, W(x)\, \varphi_k(x)\, dx \equiv \int_{\mathbb{R}^d} f_{jk}(x)\, dx$$

becomes highly oscillatory for multi-indices j and k with large components. In this situation, an estimate of the type (1.24) cannot be expected to hold true with a constant that is uniform in j and k, but rather (with $a_+ = \max\{a, 0\}$)

$$|c_m(j,k)| \le C \prod_{n=1}^{d} \bigl(1 + (m_n - j_n - k_n)_+\bigr)^{-r} \tag{1.25}$$

for the Hermite coefficients $c_m(j,k)$ of $g_{jk}(x) = f_{jk}(x)\, e^{|x|^2/2}$. This suggests a modification of Smolyak's algorithm in which terms in the sum (1.22) are discarded only if they are of size $\mathcal{O}((2^L)^{-r})$ under condition (1.25). Such an adaptation of the algorithm reads as follows: for a pair of multi-indices j and k, let $\widehat{\ell}_1, \ldots, \widehat{\ell}_d$ be such that $c \cdot 2^{\widehat{\ell}_n - 1} < \max\{j_n, k_n\} \le c \cdot 2^{\widehat{\ell}_n}$ for a chosen constant c. We discard only terms with

$$(\ell_1 - \widehat{\ell}_1)_+ + \ldots + (\ell_d - \widehat{\ell}_d)_+ > L \, .$$

In the case of a hyperbolically reduced multi-index set (1.18), we have actually

$$\widehat{\ell}_1 + \ldots + \widehat{\ell}_d \le 2 \log_2 K + \alpha d \, ,$$

where $\alpha \in \mathbb{R}$ depends only on c. Such a modification can thus be implemented by increasing L in dependence of K by $2 \log_2 K$. The number of evaluations of the potential on the resulting sparse grid thus becomes $\mathcal{O}(K^2 \cdot 2^L \cdot (L + 2\log_2 K)^{d-1})$ and hence is essentially quadratic in K of (1.18). The choice of L depends on the smoothness and growth properties of the potential.

III.1.3 Collocation Method, 1D Fourier Basis

Truncation, Periodization, Rescaling. We start from the one-dimensional Schrödinger equation (0.1) on the real line. If we expect the wavefunction to be negligible outside an interval $[a, b]$ on the considered time interval, we may replace the equation on the whole real line by that on the finite interval with periodic boundary conditions. After a rescaling and shift $x \to \alpha x + \beta$ we may assume that the space interval is $[-\pi, \pi]$:

$$i \frac{\partial \psi}{\partial t}(x,t) = -\frac{1}{2\mu} \frac{\partial^2 \psi}{\partial x^2}(x,t) + V(x)\psi(x,t) \, , \qquad x \in [-\pi, \pi] \, , \tag{1.26}$$

with periodic boundary conditions: $\psi(-\pi, t) = \psi(\pi, t)$ for all t.

Collocation by Trigonometric Polynomials. We look for an approximation to the wave function $\psi(x,t)$ by a trigonometric polynomial at every instant t,

$$\psi(x,t) \approx \psi_K(x,t) = \sum_{k=-K/2}^{K/2-1} c_k(t)\, e^{ikx} \, , \qquad x \in [-\pi, \pi] \, , \tag{1.27}$$

where K is a given even integer. We might determine the unknown Fourier coefficients $c_k(t)$ by a Galerkin method on the space of trigonometric polynomials as in the previous section. Here, we consider instead the approach by *collocation*, which requires that the approximations satisfy the Schrödinger equation in a finite number of grid points, as many points as there are unknown coefficients. We thus choose the K equidistant grid points $x_j = j \cdot 2\pi/K$ with $j = -K/2, \ldots, K/2 - 1$ and require that

$$i \frac{\partial \psi_K}{\partial t}(x_j,t) = -\frac{1}{2\mu} \frac{\partial^2 \psi_K}{\partial x^2}(x_j,t) + V(x_j)\psi(x_j,t) \quad (j = -K/2, \ldots, K/2-1). \tag{1.28}$$

This condition is equivalent to a system of ordinary differential equations for the coefficients $c_k(t)$, as we show next.

Discrete Fourier Transform. Let $\mathcal{F}_K : \mathbb{C}^K \to \mathbb{C}^K$ denote the *discrete Fourier transform* of length K, defined by

$$\widehat{v} = \mathcal{F}_K v \quad \text{with} \quad \widehat{v}_k = \frac{1}{K} \sum_{j=-K/2}^{K/2-1} e^{-ikj\cdot 2\pi/K}\, v_j \quad (k = -K/2, \dots, K/2-1). \tag{1.29}$$

The inverse transform is then $\mathcal{F}_K^{-1} = K\mathcal{F}_K^*$, that is,

$$v = \mathcal{F}_K^{-1}\widehat{v} \quad \text{with} \quad v_j = \sum_{k=-K/2}^{K/2-1} e^{ijk\cdot 2\pi/K}\, \widehat{v}_k \quad (j = -K/2, \dots, K/2-1). \tag{1.30}$$

The familiar *fast Fourier transform* (FFT) algorithm (see, e.g., the informative Wikipedia article on this topic) computes either transform with $\mathcal{O}(K \log K)$ complex multiplications and additions, instead of the K^2 operations needed for a naive direct computation from the definition.

Differential Equations for the Fourier Coefficients and Grid Values. From (1.27) we note that the vector of grid values of ψ_K is the inverse discrete Fourier transform of the coefficient vector:

$$\big(\psi_K(x_j,t)\big)_{j=-K/2}^{K/2-1} = \mathcal{F}_K^{-1}\big(c_k(t)\big)_{k=-K/2}^{K/2-1}. \tag{1.31}$$

This relation and differentiation of (1.27) yield that the collocation condition (1.28) is equivalent to the following differential equation for the vector $c = (c_k)$ of Fourier coefficients: with the diagonal matrices $D_K = \frac{1}{2\mu}\,\mathrm{diag}(k^2)$ and $V_K = \mathrm{diag}\big(V(x_j)\big)$,

$$i\dot{c} = D_K c + \mathcal{F}_K V_K \mathcal{F}_K^{-1} c\,. \tag{1.32}$$

Alternatively, by taking the inverse Fourier transform on both sides of (1.32) and recalling (1.31), we obtain a system of differential equations for the grid values $u_j(t) = \psi_K(x_j,t)$: for the vector $u = (u_j)$,

$$i\dot{u} = \mathcal{F}_K^{-1} D_K \mathcal{F}_K u + V_K u\,. \tag{1.33}$$

We observe that the matrices on the right-hand sides of (1.32) and (1.33) are all Hermitian, because $\sqrt{K}\mathcal{F}_K$ is a unitary transformation.

Approximation by Trigonometric Interpolation. For a continuous 2π-periodic function f we denote by $\mathcal{I}_K f$ the trigonometric polynomial with K Fourier modes ranging from $-K/2$ to $K/2-1$ which interpolates f in the K equidistant grid points $x_j = j \cdot 2\pi/K$:

$$\mathcal{I}_K f(x) = \sum_{k=-K/2}^{K/2-1} c_k\, e^{ikx} \quad \text{with} \quad (c_k)_{k=-K/2}^{K/2-1} = \mathcal{F}_K\big(f(x_j)\big)_{j=-K/2}^{K/2-1}.$$

Theorem 1.7 (Interpolation Error). *Suppose that f is a 2π-periodic function for which the s-th derivative $\partial_x^s f \in L^2$, for some $s \geq 1$. Then, the interpolation error is bounded in L^2 by*

$$\|f - \mathcal{I}_K f\| \leq C\, K^{-s}\, \|\partial_x^s f\|\,,$$

where C depends only on s.

Proof. We write the Fourier series of f and the trigonometric interpolation polynomial as

$$f(x) = \sum_{k=-\infty}^{\infty} a_k\, e^{ikx}\,, \qquad \mathcal{I}_K f(x) = \sum_{k=-K/2}^{K/2-1} c_k\, e^{ikx}\,.$$

From the interpolation condition it is verified that the coefficients are related by the *aliasing formula*

$$c_k = \sum_{\ell=-\infty}^{\infty} a_{k+\ell K}\,.$$

Using Parseval's formula and the Cauchy–Schwarz inequality, we thus obtain

$$\begin{aligned}
\|f - \mathcal{I}_K f\|^2 &= \sum_{k=-K/2}^{K/2-1} \Bigl(\Bigl|\sum_{\ell\neq 0} a_{k+\ell K}\Bigr|^2 + \sum_{\ell \neq 0} |a_{k+\ell K}|^2 \Bigr) \\
&\leq \sum_{k=-K/2}^{K/2-1} \Bigl(\sum_{\ell\neq 0} (k+\ell K)^{-2s} \cdot \sum_{\ell\neq 0} (k+\ell K)^{2s} |a_{k+\ell K}|^2 \\
&\qquad + \sum_{\ell\neq 0} (k+\ell K)^{-2s} \cdot (k+\ell K)^{2s} |a_{k+\ell K}|^2 \Bigr) \\
&\leq C^2\, K^{-2s} \sum_{k=-\infty}^{\infty} |k^s a_k|^2 = C^2\, K^{-2s}\, \|\partial_x^s f\|^2\,,
\end{aligned}$$

which is the desired result. □

In the same way it is shown that for every integer $m \geq 1$,

$$\|\partial_x^m (f - \mathcal{I}_K f)\| \leq C\, K^{-s}\, \|\partial_x^{s+m} f\|\,. \tag{1.34}$$

Error of the Collocation Method with Fourier Basis in 1D. We obtain the following error bound.

Theorem 1.8 (Collocation Error). *Suppose that the exact solution $\psi(t) = \psi(\cdot, t)$ has $\partial_x^{s+2}\psi(t) \in L^2$ for every $t \geq 0$, for some $s \geq 1$. Then, the error of the Fourier collocation method (1.28) with initial value $\psi_K(x,0) = \mathcal{I}_K \psi(x,0)$ is bounded in L^2 by*

$$\|\psi_K(t) - \psi(t)\| \leq C\, K^{-s}\, (1+t) \max_{0\leq\tau\leq t} \|\partial_x^{s+2}\psi(\tau)\|\,,$$

where C depends only on s.

Proof. The error analysis is based on reformulating method (1.28) as an equation with continuous argument: by interpolation on both sides of (1.28),

$$i\,\frac{\partial \psi_K}{\partial t}(x,t) = -\frac{1}{2\mu}\frac{\partial^2 \psi_K}{\partial x^2}(x,t) + \mathcal{I}_K(V\psi_K)(x,t)\,, \qquad x \in [-\pi,\pi]\,. \tag{1.35}$$

On the other hand, using that $\mathcal{I}_K V\psi = \mathcal{I}_K V \mathcal{I}_K \psi$, we obtain that the interpolant to the solution satisfies the equation

$$i\,\frac{\partial \mathcal{I}_K\psi}{\partial t}(x,t) = -\frac{1}{2\mu}\frac{\partial^2 \mathcal{I}_K\psi}{\partial x^2}(x,t) + (\mathcal{I}_K V\mathcal{I}_K\psi)(x,t) + \delta_K(x,t)\,, \tag{1.36}$$

with the defect

$$\delta_K = -\frac{1}{2\mu}\Big(\mathcal{I}_K\frac{\partial^2\psi}{\partial x^2} - \frac{\partial^2 \mathcal{I}_K\psi}{\partial x^2}\Big)\,.$$

The error $\varepsilon_K = \psi_k - \mathcal{I}_K\psi$ thus satisfies the equation

$$i\,\frac{\partial \varepsilon_K}{\partial t} = -\frac{1}{2\mu}\frac{\partial^2 \varepsilon_K}{\partial x^2} + \mathcal{I}_K(V\varepsilon_K) - \delta_K\,.$$

In terms of the Fourier coefficients $e = (e_k)$ and $d = (d_k)$ given by

$$\varepsilon_K(x,t) = \sum_{k=-K/2}^{K/2-1} e_k(t)\,e^{ikx}\,, \qquad \delta_K(x,t) = \sum_{k=-K/2}^{K/2-1} d_k(t)\,e^{ikx}\,,$$

this reads, as in (1.32):

$$i\dot e = D_K e + \mathcal{F}_K V_K \mathcal{F}_K^{-1} e - d\,,$$

with Hermitian matrices on the right-hand side, since $\mathcal{F}_K$ is unitary. Forming the Euclidean inner product with e, taking the real part and integrating we obtain, by the same argument as in the proof of Theorem II.1.5,

$$\|e(t)\| \le \|e(0)\| + \int_0^t \|d(\tau)\|\,d\tau\,.$$

By Parseval's formula, this is the same as

$$\|\varepsilon_K(t)\| \le \|\varepsilon_K(0)\| + \int_0^t \|\delta_K(\tau)\|\,d\tau\,.$$

We estimate $\delta_K(\tau)$ using Theorem 1.7 for $\partial_x^2\psi(\cdot,\tau)$ and (1.34) with $m=2$:

$$\|\delta_K(\tau)\| \le CK^{-s}\,\|\partial_x^{s+2}\psi(\cdot,\tau)\|\,.$$

Recalling that $\varepsilon_K = \psi_K - \mathcal{I}_K\psi$ and using Theorem 1.7 to estimate the interpolation error $\mathcal{I}_K\psi - \psi$, we obtain the stated result. □

Comparison with the Fourier Galerkin Method. If we use the Galerkin method (1.1) with the basis e^{-ikx} $(k = -K/2, \dots, K/2-1)$, then we obtain equations for the coefficients that are very similar to (1.32):

$$i\dot{c} = D_K c + \widehat{V}_K c\,. \tag{1.37}$$

Here, $\widehat{V}_K$ is the matrix with the entry $\frac{1}{2\pi}\int_{-\pi}^{\pi} e^{-ijx}\, V(x)\, e^{ikx}\, dx$ at position (j,k). In the collocation method (1.32), this integral is simply replaced by the trapezoidal sum approximation $\frac{1}{K}\sum_l e^{-ikx_l}\, V(x_l)\, e^{imx_l}$, with no harm to the error of the method as Theorem 1.8 shows.

III.1.4 Higher Dimensions: Hyperbolic Cross and Sparse Grids

The above results extend immediately to a full tensor-grid approximation in higher dimensions. The number of grid points and Fourier coefficients to be dealt with is then K^d in dimension d with K grid points in each direction. An approach to a reduced computational cost uses a hyperbolically reduced tensor basis of exponentials and an associated sparse grid, leading to a discretization working with $\mathcal{O}(K(\log K)^{d-1})$ grid points and Fourier coefficients. The construction is based on a discrete Fourier transform on sparse grids given by Hallatschek (1992).

Hyperbolic Cross. Instead of considering the full tensor product basis $e^{ik\cdot x} = e^{ik_1x_1}\dots e^{ik_dx_d}$ with $-K/2 \le k_n \le K/2-1$, we consider a reduced set of multi-indices $k = (k_1,\dots,k_d)$, which is constructed as follows. We order the set of integers into different levels by setting $\mathbb{Z}_0 = \{0\}$, $\mathbb{Z}_1 = \{-1\}$, $\mathbb{Z}_2 = \{-2,1\}$, $\mathbb{Z}_3 = \{-4,-3,2,3\}$, and in general

$$\mathbb{Z}_\ell = \{k \in \mathbb{Z} : -2^{\ell-1} \le k < -2^{\ell-2} \text{ or } 2^{\ell-2} \le k < 2^{\ell-1}\}\,. \tag{1.38}$$

This yields a partition of the integers into different levels as indicated in the following diagram of the line of integers:

...	-8	-7	-6	-5	-4	-3	-2	-1	0	1	2	3	4	5	6	7	...
...	4	4	4	4	3	3	2	1	0	2	3	3	4	4	4	4	...

We then define the *hyperbolic cross*

$$\begin{aligned}\mathcal{K} = \mathcal{K}_L^d = \{(k_1,\dots,k_d) :\ & \text{There are } \ell_1,\dots,\ell_d \text{ with } \ell_1+\dots+\ell_d \le L \\ & \text{such that } k_n \in \mathbb{Z}_{\ell_n} \text{ for } n=1,\dots,d\}\,.\end{aligned} \tag{1.39}$$

We will work with the basis of exponentials $e^{ik\cdot x}$ with $k \in \mathcal{K}$. As in Lemma 1.4 it is seen that $\mathcal{K}$ has $\mathcal{O}(2^L \cdot L^{d-1})$ elements.

Sparse Grid. As we now show, the wave vectors in the hyperbolic cross are in a bijective correspondence with a set of grid points in $[0,2\pi]^d$. Consider first the hierarchical ordering of grid points in the interval $[0,2\pi)$ obtained by setting $X_0 = \{0\}$, $X_1 = \{\pi\}$, $X_2 = \{\frac{\pi}{2},\frac{3\pi}{2}\}$, $X_3 = \{\frac{\pi}{4},\frac{3\pi}{4},\frac{5\pi}{4},\frac{7\pi}{4}\}$, and in general

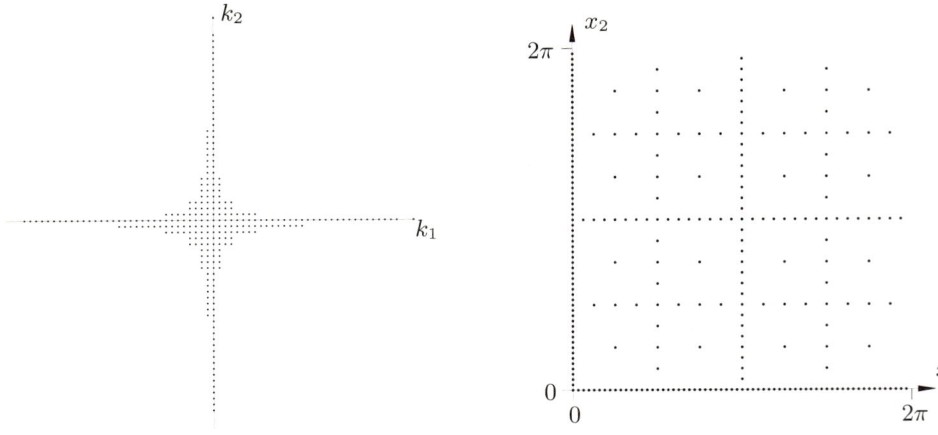

Fig. 1.4. Hyperbolic cross and sparse grid (L=6).

$$X_\ell = \left\{ (2j-1)\frac{2\pi}{2^\ell} \; : \; j = 1, \dots, 2^{\ell-1} \right\}.$$

Clearly, each grid point in X_ℓ is in a one-to-one correspondence with an integer in $\mathbb{Z}_\ell$. We define the *sparse grid* corresponding to the hyperbolic cross $\mathcal{K}$ as

$$\begin{aligned} \Gamma = \Gamma_L^d = \{(x_1, \dots, x_d) : & \text{ There are } \ell_1, \dots, \ell_d \text{ with } \ell_1 + \dots + \ell_d \le L \\ & \text{such that } x_n \in X_{\ell_n} \text{ for } n = 1, \dots, d\} . \end{aligned} \tag{1.40}$$

We will use quadrature and trigonometric interpolation on this grid.

Smolyak's Sparse-Grid Quadrature. We consider the trapezoidal (or rectangle) rule approximation to the one-dimensional integral $\frac{1}{2\pi}\int_0^{2\pi} g(x)\,dx$ of a 2π-periodic function g,

$$Q_\ell g = 2^{-\ell} \sum_{j=0}^{2^\ell - 1} g\Big(j\frac{2\pi}{2^\ell}\Big) = 2^{-\ell} \sum_{m=0}^{\ell} \sum_{x \in X_m} g(x) ,$$

and the difference between two levels,

$$\Delta_\ell g = Q_\ell g - Q_{\ell-1} g , \qquad \Delta_0 g = Q_0 g .$$

As in Section III.1.2, we consider Smolyak's quadrature for a multi-variate function $f(x_1, \dots, x_d)$, which uses values of f only on the sparse grid $\Gamma = \Gamma_L^d$:

$$S_\Gamma f = S_L^d f = \sum_{\ell_1 + \dots + \ell_d \le L} \Delta_{\ell_1} \otimes \dots \otimes \Delta_{\ell_d} f . \tag{1.41}$$

It has the following useful property.

Lemma 1.9. *Smolyak's quadrature (1.41) is exact for the exponentials $e^{ik\cdot x}$ for all multi-indices k in the hyperbolic cross $\mathcal{K}_L^d$.*

Proof. We first note that the one-dimensional trapezoidal rule Q_ℓ gives the exact value 0 for exponentials e^{ikx} whenever k is not an integral multiple of 2^ℓ, and it gives the correct value 1 for $k = 0$. With the formula

$$S_L^d f = \sum_{\ell=0}^{L} \Delta_\ell \otimes S_{L-\ell}^{d-1} f\,,$$

the result then follows by induction over the dimension. □

Remark 1.10. Unlike the full-grid case, the quadrature S_L^d is not exact for products $e^{-ijx}e^{ikx}$ with $j, k \in \mathcal{K}_L^d$. The problem arises with terms such as $j = (-2^{L-1}, 0, 0, \dots, 0)$ and $k = (0, -2^{L-1}, 0, \dots, 0)$. Since $k - j \in \mathcal{K}_{2L}^d$ for $j, k \in \mathcal{K}_L^d$, we note that such products are integrated exactly by S_{2L}^d, hence with roughly the squared number of grid points. (Cf. the similar situation in the Hermite case discussed at the end of Section III.1.2.)

Sparse-Grid Trigonometric Interpolation. The one-dimensional trigonometric interpolation of a 2π-periodic function f on a grid of 2^ℓ equidistant grid points is given as

$$I_\ell g(x) = \sum_{k=-2^{\ell-1}}^{2^{\ell-1}-1} c_k^\ell\, e^{ikx} \qquad \text{with} \quad c_k^\ell = Q_\ell(e^{-ikx} g)\,.$$

We let $\Lambda_\ell = I_\ell - I_{\ell-1}$ denote the difference operators between successive levels (with $\Lambda_0 = I_0$). The trigonometric interpolation of a multivariate function f on the full tensor grid with 2^L grid points in every coordinate direction can then be written as

$$\sum_{\ell_1=0}^{L} \dots \sum_{\ell_d=0}^{L} \Lambda_{\ell_1} \otimes \dots \otimes \Lambda_{\ell_d} f(x_1, \dots, x_d)\,.$$

Hallatschek (1992) introduces the corresponding operator with evaluations of f only on the sparse grid $\Gamma = \Gamma_L^d$ as

$$\mathcal{I}_\Gamma f(x_1, \dots, x_d) = \sum_{\ell_1+\ldots+\ell_d \le L} \Lambda_{\ell_1} \otimes \dots \otimes \Lambda_{\ell_d} f(x_1, \dots, x_d) \tag{1.42}$$

and notes the following important property.

Lemma 1.11. *$\mathcal{I}_\Gamma f$ interpolates f on the sparse grid Γ.*

Proof. This follows from the observation that the terms omitted from the full-grid interpolation operator all vanish on the sparse grid. □

Sparse Discrete Fourier Transform. We observe that $\mathcal{I}_\Gamma f(x)$ for $x = (x_1, \dots, x_d)$ is a linear combination of exponentials $e^{ik\cdot x}$ with k in the hyperbolic cross $\mathcal{K} = \mathcal{K}_L^d$:

$$\mathcal{I}_\Gamma f(x) = \sum_{k\in\mathcal{K}} c_k\, e^{ik\cdot x}\,.$$

This defines a discrete Fourier transform

$$\mathcal{F}_\Gamma : \mathbb{C}^\Gamma \to \mathbb{C}^\mathcal{K} :\ \big(f(x)\big)_{x\in\Gamma} \mapsto (c_k)_{k\in\mathcal{K}}\,. \tag{1.43}$$

With the map that determines the grid values of a trigonometric polynomial from its coefficients,

$$\mathcal{T}_\mathcal{K} : \mathbb{C}^\mathcal{K} \to \mathbb{C}^\Gamma :\ (c_k)_{k\in\mathcal{K}} \mapsto \Big(\sum_{k\in\mathcal{K}} c_k e^{ik\cdot x}\Big)_{x\in\Gamma}, \tag{1.44}$$

we have from the interpolation property that $\mathcal{T}_\mathcal{K}\mathcal{F}_\Gamma f = f$ for all $f = (f(x))_{x\in\Gamma}$, and hence $\mathcal{F}_\Gamma$ is invertible and

$$\mathcal{F}_\Gamma^{-1} = \mathcal{T}_\mathcal{K}\,. \tag{1.45}$$

Both $\mathcal{F}_\Gamma$ and its inverse can be implemented with $\mathcal{O}(2^L \cdot L^d)$ operations, using one-dimensional FFTs and hierarchical bases; see Hallatschek (1992) and Gradinaru (2007).

There is no discrete Parseval formula for $\mathcal{F}_\Gamma$, but by Remark 1.10, the following restricted Parseval relation is still valid: with the inner product $\langle f \,|\, g\rangle_\Gamma = S_\Gamma(\overline{f}g)$ on Γ and the Euclidean inner product $\langle \cdot \,|\, \cdot\rangle_\mathcal{K}$ on $\mathcal{K}$,

$$\langle \mathcal{F}_\Gamma^{-1} c \,|\, \mathcal{F}_\Gamma^{-1} d\rangle_\Gamma = \langle c \,|\, d\rangle_\mathcal{K} \qquad \text{if } c_k = d_k = 0 \text{ for } k \in \mathcal{K}_L^d \setminus \mathcal{K}_{L/2}^d\,. \tag{1.46}$$

Approximation by Sparse-Grid Trigonometric Interpolation. Error bounds are given by Hallatschek (1992) in the maximum norm, and by Gradinaru (2008) in L^2 and related norms. The L^2 error bound reads

$$\|\mathcal{I}_\Gamma f - f\| \le C(d,s)\,(L+1)^{d-1}\,(2^L)^{-s}\,\|\partial_{x_1}^{s+1}\ldots\partial_{x_d}^{s+1} f\|\,. \tag{1.47}$$

The estimate is obtained by carefully estimating the terms $\Lambda_{\ell_1}\otimes\ldots\otimes\Lambda_{\ell_d} f$ that have been omitted in (1.42).

Collocation of the Schrödinger Equation on Sparse Grids. Gradinaru (2008) studies the collocation method, which approximates the solution by a trigonometric polynomial with coefficients on the hyperbolic cross,

$$\psi_\mathcal{K}(x,t) = \sum_{k\in\mathcal{K}} c_k(t)\, e^{ik\cdot x}\,, \tag{1.48}$$

and requires the Schrödinger equation to hold in the points of the sparse grid. This yields the system for the Fourier coefficients $c = (c_k)_{k\in\mathcal{K}}$,

$$i\dot{c} = D_\mathcal{K} c + \mathcal{F}_\Gamma V_\Gamma \mathcal{F}_\Gamma^{-1} c\,, \tag{1.49}$$

where $(D_\mathcal{K} c)_k = \frac{1}{2\mu}\,|k|^2 c_k$ for $k\in\mathcal{K}$, and V_Γ is the diagonal matrix with entries $V(x)$ for $x\in\Gamma$. Gradinaru (2008) shows that the error of the collocation method over bounded time intervals is bounded by $\mathcal{O}(L^{d-1}\,(2^L)^{-s})$ if mixed derivatives up to order $s+2$ in each coordinate direction are bounded in L^2.

An unpleasant feature in (1.49) is the fact that the matrix $\mathcal{F}_\Gamma V_\Gamma \mathcal{F}_\Gamma^{-1}$ is not Hermitian, since the sparse-grid Fourier transform $\mathcal{F}_\Gamma$ is not a scalar multiple of a unitary operator, unlike the full tensor-grid case. This can give numerical artefacts such as the loss of conservation of norm and in theory may lead to an exponential, instead of linear, error growth in time, with a rate that is given by a bound of the skew-Hermitian part of $\mathcal{F}_\Gamma V_\Gamma \mathcal{F}_\Gamma^{-1}$. Moreover, some of the time-stepping methods considered in the subsequent sections are not applicable in the case of non-Hermitian matrices.

Discretizations on Sparse Grids Having Hermitian Matrices. Are there methods with similar complexity and approximation properties to the sparse-grid collocation method but which have a Hermitian matrix? We start from the interpretation of the collocation method as a Galerkin method with trapezoidal rule approximation of the integrals in the matrix elements, as noted at the end of Section III.1.3, and aim for a multi-dimensional, sparse-grid extension that approximates the matrix elements by Smolyak's quadrature.

We consider the inner product on $\mathbb{C}^\Gamma$ defined by Smolyak's quadrature on the sparse grid,

$$\langle f\,|\,g\rangle_\Gamma = S_\Gamma(\overline{f}g)\,,$$

and the Euclidean inner product $\langle\cdot\,|\,\cdot\rangle_\mathcal{K}$ on $\mathbb{C}^\mathcal{K}$. With respect to these inner products, we take the adjoint $(\mathcal{F}_\Gamma^{-1})^*$ of $\mathcal{F}_\Gamma^{-1}$:

$$\langle \mathcal{F}_\Gamma^{-1} a\,|\,f\rangle_\Gamma = \langle a\,|\,(\mathcal{F}_\Gamma^{-1})^* f\rangle_\mathcal{K} \qquad \forall\, f\in\mathbb{C}^\Gamma,\ a\in\mathbb{C}^\mathcal{K}.$$

Then, $(\mathcal{F}_\Gamma^{-1})^* f = \big(S_\Gamma(e^{-ik\cdot x} f)\big)_{k\in\mathcal{K}}$, and we obtain that

$$(\mathcal{F}_\Gamma^{-1})^* V_\Gamma \mathcal{F}_\Gamma^{-1} = \Big(S_\Gamma\big(e^{-ij\cdot x}\,V(x)\,e^{ik\cdot x}\big)\Big)_{j,k\in\mathcal{K}}$$

is the Hermitian matrix that contains the sparse-grid quadrature approximations to the Galerkin matrix elements.

Instead of (1.49) we would like to determine the coefficients of (1.48) from

$$i\dot{c} = D_\mathcal{K} c + (\mathcal{F}_\Gamma^{-1})^* V_\Gamma \mathcal{F}_\Gamma^{-1} c\,. \tag{1.50}$$

This method can be rewritten as a quasi-Galerkin method on the hyperbolic-cross space $\mathcal{V}_\mathcal{K} = \mathrm{span}\{e^{ik\cdot x} : k\in\mathcal{K}\}$: determine $\psi_\mathcal{K}(t)\in\mathcal{V}_\mathcal{K}$ (i.e., of the form (1.48)) such that

$$\Big\langle \varphi_\mathcal{K}\,\Big|\, i\,\frac{\partial\psi_\mathcal{K}}{\partial t}\Big\rangle = \Big\langle \varphi_\mathcal{K}\,\Big|\, -\frac{1}{2\mu}\Delta\psi_\mathcal{K}\Big\rangle + \Big\langle \varphi_\mathcal{K}\,\Big|\, V\psi_\mathcal{K}\Big\rangle_\Gamma \qquad \forall\,\varphi_\mathcal{K}\in\mathcal{V}_\mathcal{K}\,. \tag{1.51}$$

Here, the last inner product is the discrete inner product on the sparse grid instead of the usual L^2 inner product. Unfortunately, it appears that this does *not* give a convergent discretization for the hyperbolic cross $\mathcal{K}=\mathcal{K}_L^d$ and the sparse grid $\Gamma=\Gamma_L^d$ of the same level L. We describe three ways to cope with this difficulty:

1. *Discrete Galerkin Method with a Simplified Mass Matrix:* We replace the L^2 inner products in (1.51) by the discrete inner product on $\Gamma = \Gamma_L^d$. Then we obtain a standard Galerkin method with a discrete inner product. The associated orthogonal projection to $\mathcal{V}_\mathcal{K}$ is just the interpolation $\mathcal{I}_\Gamma$. Optimal error bounds are then obtained with the standard proof for Galerkin methods, as in Theorem 1.3. However, since the exponentials $e^{ik\cdot x}$, $k \in \mathcal{K}$, do not form an orthonormal basis with respect to the discrete inner product, there are now non-diagonal matrices

$$M_\mathcal{K} = \big(m_{jk}\big)_{j,k\in\mathcal{K}} = \big(\langle e^{ij\cdot x} \,|\, e^{ik\cdot x}\rangle_\Gamma\big)_{j,k\in\mathcal{K}}, \qquad T_\mathcal{K} = \frac{1}{2\mu}\big(j\cdot k\, m_{jk}\big)_{j,k\in\mathcal{K}}$$

in the differential equations for the coefficients:

$$M_\mathcal{K}\dot{c} = T_\mathcal{K} c + (\mathcal{F}_\Gamma^{-1})^* V_\Gamma \mathcal{F}_\Gamma^{-1} c\,.$$

By (1.46), the mass matrix partitioned into blocks corresponding to $\mathcal{K}_{L/2}^d$ and $\mathcal{K}_L^d \setminus \mathcal{K}_{L/2}^d$ takes the form

$$M_\mathcal{K} = \begin{pmatrix} I & B^* \\ B & N \end{pmatrix}$$

with sparse matrices B and N. An approximate Choleski factor of $M_\mathcal{K}$ is given by

$$C = \begin{pmatrix} I & 0 \\ B & I \end{pmatrix} \quad \text{with} \quad C^{-1} = \begin{pmatrix} I & 0 \\ -B & I \end{pmatrix} \quad \text{and} \quad CC^* = \begin{pmatrix} I & B^* \\ B & I + BB^* \end{pmatrix},$$

where only the lower diagonal block differs from that in $M_\mathcal{K}$. Replacing $M_\mathcal{K}$ by CC^*, we obtain for $b = C^*c$

$$\dot{b} = C^{-1}T_\mathcal{K}(C^{-1})^*b + C^{-1}(\mathcal{F}_\Gamma^{-1})^* V_\Gamma \mathcal{F}_\Gamma^{-1}(C^{-1})^*b\,. \tag{1.52}$$

Since only the lower diagonal block of $M_\mathcal{K}$ has been changed, we can still get error bounds as for the full Galerkin method, but with 2^{-L} replaced by $2^{-L/2}$.

2. *Discrete Galerkin Method with Refined Sparse Grid.* By Lemma 1.9, the mass matrix becomes the identity matrix if we choose the finer grid

$$\Gamma = \Gamma_{2L}^d$$

with $2L$ instead of L levels and thus, alas, roughly the squared number of grid points. In that case, the L^2 inner products (1.51) are equal to the discrete inner products on Γ, and we obtain a standard Galerkin method with a discrete inner product. The associated orthogonal projection to $\mathcal{V}_\mathcal{K}$ is $P_\mathcal{K}\mathcal{I}_\Gamma$, where $P_\mathcal{K}$ is the orthogonal projection with respect to the L^2 inner product. Optimal error bounds are then obtained with the standard proof for Galerkin methods, as in Theorem 1.3.

3. *Galerkin Method with an Approximated Potential.* We use the standard Galerkin method with L^2 inner products, and compute the matrix elements of the potential, $\langle e^{ij\cdot x} \,|\, V \,|\, e^{ik\cdot x}\rangle$, *exactly* for an approximated potential $V(x) \approx \sum_{m\in\mathcal{M}} v_m e^{im\cdot x}$ (possibly over a coarser hyperbolic cross $\mathcal{M} \subset \mathcal{K}$), noting that $\langle e^{ij\cdot x} \,|\, e^{im\cdot x}e^{ik\cdot x}\rangle \neq 0$ only for $j = k + m$. This requires $\mathcal{O}(\#\mathcal{M} \cdot \#\mathcal{K})$ operations for computing a matrix-vector product.

III.2 Polynomial Approximations to the Matrix Exponential

After space discretization, we are left with a linear system of differential equations

$$i\dot{y} = Ay \tag{2.1}$$

with a Hermitian matrix A of large dimension and of large norm, such as (1.3) or (1.32) or (1.50) or (1.52). The latter example also shows how to deal with the presence of a mass matrix by a (possibly incomplete) Choleski decomposition. The solution to the initial value $y(0) = y_0$ is given by the matrix exponential

$$y(t) = e^{-itA} y_0 \,. \tag{2.2}$$

We study time stepping methods that advance the approximate solution[1] from time t^n to $t^{n+1} = t^n + \Delta t$, from y^n to y^{n+1}. In the present section we consider methods that require only multiplications of the matrix A with vectors, and hence are given by polynomial approximations $P(\Delta t A)$ to the exponential:

$$y^{n+1} = P(\Delta t A)\, y^n \,.$$

We consider in detail the *Chebyshev method*, where the polynomial is chosen *a priori* from given information on the extreme eigenvalues of A, and the *Lanczos method*, where the polynomial is determined by a Galerkin method on the Krylov subspace, which consists of the products of all polynomials of $\Delta t A$ of a given degree with the starting vector. In the Lanczos method, a different polynomial is implicitly selected in every time step.

We mention in passing that there are further interesting methods that require only matrix-vector products with A: the *Leja point method* has similar approximation properties to the Chebyshev method but in contrast to the Chebyshev method, higher-degree polynomials of the family are constructed by reusing the computations for the lower-degree polynomials, cf. Caliari, Vianello & Bergamaschi (2004); explicit *symplectic methods* preserve the symplectic structure of the differential equation, see Gray & Manolopoulos (1996) and Blanes, Casas & Murua (2006).

III.2.1 Chebyshev Method

A near-optimal polynomial approximation to the exponential is given by its truncated Chebyshev expansion. We describe this approach, which in the context of Schrödinger equations was put forward by Tal-Ezer & Kosloff (1984), and give an error analysis based on Bernstein's theorem on polynomial approximations to analytic functions on an interval. We refer to Rivlin (1990) for background information on Chebyshev polynomials and to Markushevich (1977), Chap. III.3, for the polynomial approximation theory based on Faber polynomials.

[1] The time step number n will always be indicated as superscript in the notation.

Chebyshev Polynomials. For every non-negative integer k, the function defined by

$$T_k(x) = \cos(k\theta) \qquad \text{with} \quad \theta = \arccos x \in [0, \pi], \quad \text{for } x \in [-1, 1] \tag{2.3}$$

is in fact a polynomial of degree k, named the kth *Chebyshev polynomial*. This fact is seen from the recurrence relation

$$T_{k+1}(x) = 2xT_k(x) - T_{k-1}(x)\,, \qquad k \ge 1\,, \tag{2.4}$$

starting from $T_0(x) = 1$ and $T_1(x) = x$, which is obtained from the trigonometric identity $\cos((n+1)\theta) + \cos((n-1)\theta) = 2\cos\theta\,\cos(n\theta)$. The Chebyshev polynomials are orthogonal polynomials with respect to the weight function $(1-x^2)^{-1/2}$ on $[-1,1]$:

$$\int_{-1}^{1} T_j(x)\,T_k(x)\,\frac{dx}{\sqrt{1-x^2}} = 0 \qquad \text{for} \quad j \ne k\,, \tag{2.5}$$

as is seen by substituting $x = \cos\theta$ and $dx/\sqrt{1-x^2} = d\theta$ and using the orthogonality of the complex exponentials.

Another useful formula is

$$2\,T_k(x) = \bigl(x + \sqrt{x^2-1}\bigr)^k + \bigl(x - \sqrt{x^2-1}\bigr)^k, \tag{2.6}$$

again verified by substituting $x = \cos\theta$. The *Joukowski transform*

$$w = \Phi(z) = z + \sqrt{z^2-1}\,, \qquad z = \Psi(w) = \frac{1}{2}\Bigl(w + \frac{1}{w}\Bigr) \tag{2.7}$$

is the conformal map between the exterior of the interval $[-1,1]$ and the exterior of the unit disk, $|w| > 1$. (The branch of the square root is chosen such that $\sqrt{z^2-1} \sim z$ for $z \to \infty$.) The level sets $\Gamma_r = \{z : |\Phi(z)| = r\} = \{\Psi(w) : |w| = r\}$ for $r > 1$ are ellipses with foci ± 1, major semi-axis $r + r^{-1}$ and minor semi-axis $r - r^{-1}$. Since the Laurent expansion at ∞ of $(z - \sqrt{z^2-1})^k$ contains only powers z^{-j} with $j \ge k$, the integral of that function over a closed contour Γ encircling the interval $[-1,1]$ vanishes by Cauchy's theorem. With Cauchy's integral formula we thus obtain from (2.6)

$$2\,T_k(x) = \frac{1}{2\pi i}\int_\Gamma \frac{\Phi(z)^k}{z-x}\,dz\,, \qquad x \in [-1,1], \tag{2.8}$$

which establishes an important relationship between the Chebyshev polynomials and the conformal map: the Chebyshev polynomials are the *Faber polynomials* for the interval $[-1,1]$; cf. Markushevich (1977), Sect. III.3.14.

Chebyshev and Fourier Series. Given a (smooth) complex-valued function $f(x)$ on the interval $-1 \le x \le 1$, we expand the 2π-periodic, symmetric function

$$g(\theta) = f(\cos\theta)$$

as a Fourier series:

$$g(\theta) = \sum_{k=-\infty}^{\infty} c_k \, e^{ik\theta} \qquad \text{with} \quad c_k = \frac{1}{2\pi} \int_{-\pi}^{\pi} e^{-ik\theta} \, g(\theta) \, d\theta$$

or in fact, by the symmetry $g(-\theta) = g(\theta)$,

$$g(\theta) = c_0 + 2 \sum_{k=1}^{\infty} c_k \cos(k\theta) \qquad \text{with} \quad c_k = \frac{1}{\pi} \int_0^{\pi} \cos(k\theta) \, g(\theta) \, d\theta \, .$$

Substituting $x = \cos\theta$ and $dx/\sqrt{1-x^2} = d\theta$, we obtain the *Chebyshev expansion*

$$f(x) = c_0 + 2 \sum_{k=1}^{\infty} c_k \, T_k(x) \qquad \text{with} \quad c_k = \frac{1}{\pi} \int_{-1}^{1} T_k(x) \, f(x) \, \frac{dx}{\sqrt{1-x^2}} \, . \tag{2.9}$$

Chebyshev Approximation of Holomorphic Functions. We study the approximation of a holomorphic function $f(x)$ by the truncated series with m terms,

$$S_m f(x) = c_0 + 2 \sum_{k=1}^{m-1} c_k \, T_k(x) \, ,$$

which is a polynomial of degree $m-1$. The following is a version of a theorem by Bernstein (1912); see Markushevich (1977), Sect. III.3.15. Here $\Phi(z) = z + \sqrt{z^2-1}$ is again the conformal map (2.7) from the complement of the interval $[-1, 1]$ to the exterior of the unit disk, and $\Psi(w) = \frac{1}{2}(w + \frac{1}{w})$ is the inverse map.

Theorem 2.1 (Chebyshev Approximation). *Let $r > 1$, and suppose that $f(z)$ is holomorphic in the interior of the ellipse $|\Phi(z)| < r$ and continuous on the closure. Then, the error of the truncated Chebyshev series is bounded by*

$$|f(x) - S_m f(x)| \le 2 \, \mu(f, r) \, \frac{r^{-m}}{1 - r^{-1}} \qquad \text{for} \quad -1 \le x \le 1,$$

with the mean value $\mu(f, r) = \frac{1}{2\pi r} \int_{|w|=r} |f(\Psi(w))| \cdot |dw|$.

Proof. We start from the Cauchy integral formula over the ellipse $\Gamma_r = \{z : |\Phi(z)| = r\} = \{\Psi(w) : |w| = r\}$ and substitute $z = \Psi(w)$:

$$f(x) = \frac{1}{2\pi i} \int_{\Gamma_r} \frac{f(z)}{z - x} \, dz = \frac{1}{2\pi i} \int_{|w|=r} f(\Psi(w)) \, \frac{\Psi'(w)}{\Psi(w) - x} \, dw \, . \tag{2.10}$$

We expand in negative powers of w,

$$\frac{\Psi'(w)}{\Psi(w) - x} = \sum_{k=0}^{\infty} a_k(x) \, w^{-k-1} \qquad \text{for} \quad |w| > 1, \tag{2.11}$$

where the Taylor coefficients at ∞ are given as

$$a_k(x) = \frac{1}{2\pi i}\int_{|w|=r} w^k \,\frac{\Psi'(w)}{\Psi(w)-x}\,dw = \frac{1}{2\pi i}\int_{\Gamma_r} \frac{\Phi(z)^k}{z-x}\,dz\,.$$

By (2.8), these coefficients turn out to be simply

$$a_k(x) = 2\,T_k(x).$$

Inserting (2.11) into (2.10) therefore yields

$$f(x) - S_m f(x) = \frac{1}{2\pi i}\int_{|w|=r} f(\Psi(w))\cdot 2\sum_{k=m}^{\infty} T_k(x)\,w^{-k-1}\,dw\,.$$

Since $|T_k(x)| \le 1$ for $-1 \le x \le 1$, we have for $|w| = r > 1$

$$\Bigl|\sum_{k=m}^{\infty} T_k(x)\,w^{-k-1}\Bigr| \le \sum_{k=m}^{\infty} r^{-k-1} = \frac{r^{-m-1}}{1-r^{-1}}\,,$$

and the result follows. □

Chebyshev Approximation of Complex Exponentials. The complex exponential $e^{i\omega x}$ is an entire function, and we can choose r in Theorem 2.1 dependent on m to balance the growth of $\mu(e^{i\omega z}, r)$ with r against the decay of r^{-m}. This gives the following corollary showing superlinear convergence after a stagnation up to $m \approx |\omega|$. Since the polynomial must capture the extrema and zeros of $\cos(\omega x)$ and $\sin(\omega x)$ for a uniform approximation, it is obvious that at least a degree m proportional to $|\omega|$ is needed to obtain an error uniformly smaller than 1. Once this barrier is surmounted, the error decays very rapidly with growing degree m.

Theorem 2.2 (Eventual Superlinear Convergence to $\mathrm{e}^{i\omega x}$). *The error of the Chebyshev approximation $p_{m-1}(x)$ of degree $m-1$ to the complex exponential $e^{i\omega x}$ with real ω is bounded by*

$$\max_{-1\le x\le 1} |p_{m-1}(x) - e^{i\omega x}| \le 4\left(e^{1-(\omega/2m)^2}\,\frac{|\omega|}{2m}\right)^m \quad \text{for} \quad m \ge |\omega|\,. \qquad (2.12)$$

Proof. We have $\mu(e^{i\omega z}, r) \le \max_{z\in\Gamma_r} |e^{i\omega z}| = e^{|\omega|(r-r^{-1})/2}$, where the maximum is attained at $z = \pm\frac{1}{2}(ir + \frac{1}{ir})$ on the minor semi-axis. Theorem 2.1 thus gives us the bound

$$\max_{-1\le x\le 1} |p_{m-1}(x) - e^{i\omega x}| \le \frac{2r^{-m}}{1-r^{-1}}\, e^{|\omega|(r-r^{-1})/2}\,.$$

Choosing $r = 2m/|\omega| \ge 2$ then yields the stated result, which could be slightly refined. □

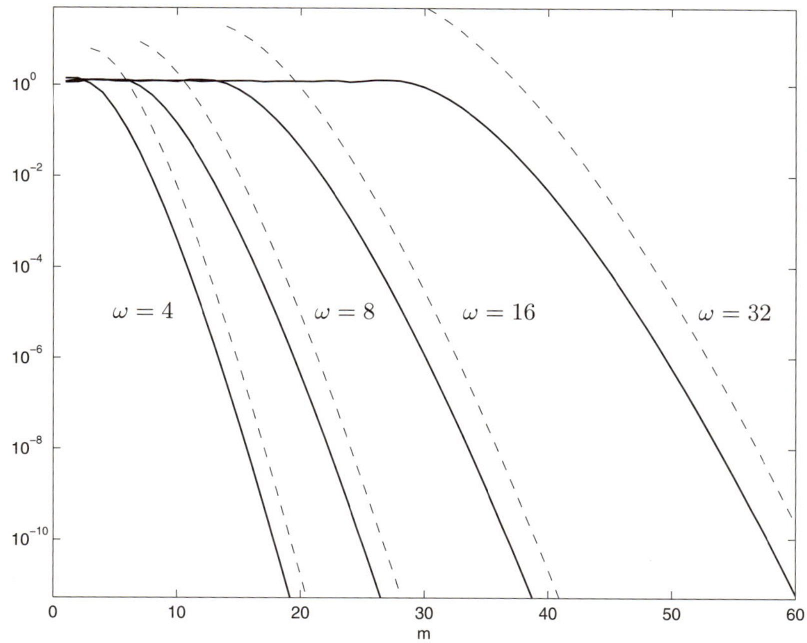

Fig. 2.1. Chebyshev approximation of $e^{i\omega x}$. Maximum error on $[-1,1]$ versus degree, for $\omega = 4, 8, 16, 32$. Dashed: Error bounds of Theorem 2.2.

The Chebyshev coefficients of $e^{i\omega x}$ are given explicitly by Bessel functions of the first kind: by formula (9.1.21) in Abramowitz & Stegun (1965),

$$c_k = \frac{1}{\pi}\int_0^\pi e^{i\omega\cos\theta}\cos(k\theta)\,d\theta = i^k J_k(\omega)\,. \tag{2.13}$$

Transforming the Interval. From $e^{i\omega x}$ with $-1 \le x \le 1$, uniform polynomial approximation of $e^{-i\xi}$ for $\alpha \le \xi \le \beta$ is obtained by transforming

$$x = \frac{2}{\beta-\alpha}\Big(\xi - \frac{\alpha+\beta}{2}\Big), \qquad \xi = \frac{\alpha+\beta}{2} + x\,\frac{\beta-\alpha}{2}\,.$$

We then approximate $e^{-i\xi} = e^{-i(\alpha+\beta)/2}\,e^{-ix(\beta-\alpha)/2}$ using $e^{-ix(\beta-\alpha)/2} \approx c_0 + 2\sum_{k=1}^{m-1} c_k T_k(x)$ with $c_k = i^k J_k\big(-\frac{\beta-\alpha}{2}\big) = (-i)^k J_k\big(\frac{\beta-\alpha}{2}\big)$, so that

$$e^{-i\xi} \approx e^{-i(\alpha+\beta)/2}\left(c_0 + 2\sum_{k=1}^{m-1} c_k\, T_k\Big(\frac{2}{\beta-\alpha}\Big(\xi - \frac{\alpha+\beta}{2}\Big)\Big)\right) \qquad \text{for} \quad \alpha \le \xi \le \beta\,.$$

Chebyshev Method for the Matrix Exponential Operator. Let A be a Hermitian matrix all of whose eigenvalues are known to lie in the interval $[a,b]$. As proposed by Tal-Ezer & Kosloff (1984), we approximate the action of the matrix exponential on a vector v by

$$e^{-i\Delta t A} v \approx P_{m-1}(\Delta t\, A)v\,,$$

where

$$P_{m-1}(\Delta t\, A)v = e^{-i\Delta t(a+b)/2}\left(c_0 v + 2\sum_{k=1}^{m-1} c_k\, T_k\Big(\frac{2}{(b-a)}\Big(A - \frac{(a+b)}{2}I\Big)\Big)v\right) \tag{2.14}$$

with $c_k = (-i)^k J_k(\Delta t(b-a)/2)$. We observe that the right-hand side is in fact a function of the product $\Delta t\, A$. The actual way to compute (2.14) is by a recursive algorithm proposed by Clenshaw (1962) for the evaluation of truncated Chebyshev expansions of functions.

Algorithm 2.3 (Clenshaw Algorithm). *Let* $X = \frac{2}{(b-a)}\Big(A - \frac{(a+b)}{2}I\Big)$, *initialize* $d_{m+1} = d_m = 0$ *and compute recursively*

$$d_k = c_k v + 2X d_{k+1} - d_{k+2} \quad \textit{for} \quad k = m-1, m-2, \dots, 0\,.$$

Then, the approximation (2.14) is given as

$$P_{m-1}(\Delta t\, A)v = d_0 - d_2\,.$$

This identity is readily verified using the Chebyshev recurrence relation (2.4) for the terms in the sum, descending from the terms of highest degree. The algorithm requires m matrix-vector multiplications to compute $P_{m-1}(\Delta t\, A)v$ and needs to keep only three vectors in memory.

Theorem 2.4 (Error of the Chebyshev Method). *Let A be a Hermitian matrix with all its eigenvalues in the interval $[a,b]$, and let v be a vector of unit Euclidean norm. Then, the error of the Chebyshev approximation (2.14) is bounded in the Euclidean norm by*

$$\|P_{m-1}(\Delta t\, A)v - e^{-i\Delta t A}v\| \le 4\left(e^{1-(\omega/2m)^2}\,\frac{\omega}{2m}\right)^m \quad \textit{for} \quad m \ge \omega$$

with $\omega = \Delta t\,(b-a)/2$.

Proof. For a diagonal matrix A, the estimate follows immediately from Theorem 2.2 and the linear transformation between the intervals $[\Delta t\, a, \Delta t\, b]$ and $[-1,1]$. Since every Hermitian matrix A can be unitarily transformed to diagonal form, we obtain the result as stated. □

Step Size Restriction. The condition $m \ge \omega$ can be read as a restriction of the step size for given degree m:

$$\Delta t \le \frac{2m}{b-a}\,.$$

This can also be viewed as saying that at least one matrix-vector multiplication is needed on every time interval of length $1/(b-a)$. In the treatment of the Schrödinger equation, this length shrinks as the spatial discretization is refined: for illustration, consider Fourier collocation in one space dimension, with K Fourier modes. For the matrix $A = D_K + \mathcal{F}_K V_K \mathcal{F}_K^{-1}$ of (1.32), the eigenvalues lie in the interval $[a,b]$ with

$$a = \min_x V(x)\,, \qquad b = \frac{1}{2\mu}\frac{K^2}{4} + \max_x V(x)\,.$$

For large K, or small $\Delta x = 2\pi/K$, we have that $\omega = \Delta t(b-a)/2$ is approximately proportional to $\Delta t\, K^2$, or

$$\omega \sim \frac{\Delta t}{\Delta x^2}\,.$$

The condition $m \ge \omega$ for the onset of error reduction therefore translates into a step-size restriction

$$\Delta t \le C\, m\, \Delta x^2\,, \tag{2.15}$$

and the number of matrix-vector multiplications to cover a given time interval is thus inversely proportional to Δx^2.

III.2.2 Lanczos Method

A different approach to approximately computing $e^{-i\Delta t A}v$ using only the action of A on vectors is based on a Galerkin approximation to $i\dot{y} = Ay$ on the Krylov space spanned by $v, Av, \dots, A^{m-1}v$. A suitable basis for this space is given by the Lanczos iteration, named after Lanczos (1950), which has become a classic in numerical linear algebra primarily because of its use for eigenvalue problems and solving linear systems; see, e.g., Golub & Van Loan (1996), Chap. 9, and Trefethen & Bau (1997), Chap. VI. The use of the Lanczos method for approximating $e^{-i\Delta t A}v$ was first proposed by Park & Light (1986), properly in the context of approximating the evolution operator of the Schrödinger equation. Krylov subspace approximation to the matrix exponential operator has since been found useful in a variety of application areas — and has been honourably included as the twentieth of the "Nineteen dubious ways to compute the exponential of a matrix" by Moler & Van Loan (2003). Error analyses, both for the Hermitian and non-Hermitian case, have been given by Druskin & Knizhnerman (1995), Hochbruck & Lubich (1997), and Saad (1992).

Krylov Subspace and Lanczos Basis. Let A be an $N \times N$ Hermitian matrix, and let v be a non-zero complex N-vector. The mth *Krylov subspace* of $\mathbb{C}^N$ with respect to A and v is

$$\mathcal{K}_m(A,v) = \mathrm{span}(v, Av, A^2v, \dots, A^{m-1}v)\,, \tag{2.16}$$

that is, the space of all polynomials of A up to degree $m-1$ acting on the vector v.

The *Hermitian Lanczos method* builds an orthonormal basis of this space by Gram-Schmidt orthogonalization: beginning with $v_1 = v/\|v\|$, it constructs v_{k+1} recursively for $k = 1, 2, \dots$ by orthogonalizing Av_k against the previous v_j and normalizing:

$$\tau_{k+1,k}\, v_{k+1} = Av_k - \sum_{j=1}^{k} \tau_{jk}\, v_j \tag{2.17}$$

with $\tau_{jk} = v_j^* A v_k$ for $j \le k$, and with $\tau_{k+1,k} > 0$ determined such that v_{k+1} is of unit Euclidean norm — unless the right-hand side is zero, in which case the dimension of $\mathcal{K}_m(A, v)$ is k for $m \ge k$ and the process terminates.

By the mth step, the method generates the $N \times m$ matrix $V_m = (v_1 \dots v_m)$ having the orthonormal Lanczos vectors v_k as columns, and the $m \times m$ matrix $T_m = (\tau_{jk})$ with $\tau_{jk} = 0$ for $j - k > 1$. Because of (2.17), these matrices are related by

$$AV_m = V_m T_m + \tau_{m+1,m} v_{m+1} e_m^T, \tag{2.18}$$

where $e_m^T = (0 \dots 0\, 1)$ is the mth unit vector. By the orthonormality of the Lanczos vectors v_k, this equation implies

$$T_m = V_m^* A V_m \,, \tag{2.19}$$

which shows in particular that T_m is a Hermitian matrix, and hence a tridiagonal matrix: $\tau_{jk} = 0$ for $|j - k| > 1$. The sum in (2.17) therefore actually contains only the two terms for $j = k-1, k$. For a careful practical implementation, error propagation and the loss of orthogonality due to rounding errors are a concern for larger m, and (selective) reorthogonalization can substantially improve the stability properties; see Golub & Van Loan (1996), Sect. 9.2. The following is a standard version of the Lanczos iteration without reorthogonalization of the Lanczos vectors.

Algorithm 2.5 (Hermitian Lanczos Algorithm Without Reorthogonalization). *Given a Hermitian matrix A and a vector v of unit norm, the algorithm computes the Lanczos vectors $v_1, \dots, v_m$ and the entries $\alpha_j = \tau_{j,j}$ and $\beta_{j+1} = \tau_{j+1,j}$ of the tridiagonal matrix T_m. After initializing $v_1 := v$, $v_0 := 0$, $\beta_1 := 0$, the Lanczos iteration runs as follows, for $j = 1, \dots, m$:*

$$\begin{aligned} u &:= Av_j - \beta_j v_{j-1}\,, & \alpha_j &:= \langle v_j \,|\, u \rangle \\ u &:= u - \alpha_j v_j\,, & \beta_{j+1} &:= \|u\| \\ v_{j+1} &:= u/\beta_{j+1}\,. \end{aligned}$$

Galerkin Method on the Krylov Subspace. Following Park & Light (1986), we consider the Galerkin method (1.1) for the approximation of the initial value problem

$$i\dot{y} = Ay\,, \quad y(0) = v \quad \text{with} \quad \|v\| = 1$$

on the Krylov subspace $\mathcal{K}_m(A,v)$ with $m \ll N$ ($m \le 20$, say): we determine an approximation $u_m(t) \in \mathcal{K}_m(A,v)$ with $u_m(0) = v$ such that at every instant t, the time derivative satisfies

$$\langle w_m \,|\, i\dot{u}_m(t) - Au_m(t)\rangle = 0 \qquad \forall\, w_m \in \mathcal{K}_m(A,v)\,.$$

Writing $u_m(t)$ in the Lanczos basis,

$$u_m(t) = \sum_{k=1}^{m} c_k(t)\, v_k = V_m c(t) \qquad \text{with} \quad c(t) = \big(c_k(t)\big)\,,$$

we obtain for the coefficients the linear differential equation

$$i\dot{c}(t) = T_m c(t)\,, \qquad c(0) = e_1 = (1,0,\dots,0)^T$$

with the Lanczos matrix $T_m = (v_j^* A v_k)_{j,k=1}^m$ of (2.19). Clearly, the solution is given by $c(t) = e^{-itT_m} e_1$. The Galerkin approximation $u_m(t) = V_m c(t)$ at time Δt is thus the result of the following algorithm.

Algorithm 2.6 (Lanczos Method for the Exponential). *With the Lanczos matrices V_m and T_m, approximate*

$$e^{-i\Delta t A} v \approx V_m\, e^{-i\Delta t T_m}\, e_1\,. \tag{2.20}$$

For the small tridiagonal Hermitian matrix T_m, the exponential is readily computed from a diagonalization of T_m. The algorithm needs to keep all the Lanczos vectors in memory, which may not be feasible for large problems. In such a situation, the Lanczos iteration may be run twice with only four vectors in memory: in a first run for computing T_m, and in a second run (without recomputing the already known inner products) for forming the linear combination of the Lanczos vectors according to (2.20).

By the interpretation of (2.20) as a Galerkin method, we know from Sect. II.1 that norm and energy are preserved.

A Posteriori Error Bound and Stopping Criterion. From Theorem II.1.5 with the Krylov subspace as approximation space we have the error bound

$$\|u_m(t) - y(t)\| \le \int_0^t \mathrm{dist}\big(Au_m(s), \mathcal{K}_m(A,v)\big)\, ds\,.$$

By (2.18) we have

$$Au_m(s) = AV_m\, e^{-isT_m}\, e_1 = V_m T_m\, e^{-isT_m}\, e_1 + \tau_{m+1,m}\, v_{m+1}\, e_m^T\, e^{-isT_m}\, e_1$$

and therefore

$$\mathrm{dist}\big(Au_m(s), \mathcal{K}_m(A,v)\big) = \tau_{m+1,m} \left|\big[e^{-isT_m}\big]_{m,1}\right|,$$

where $[\cdot]_{m.1}$ denotes the $(m,1)$ element of a matrix. This gives us the following computable error bound.

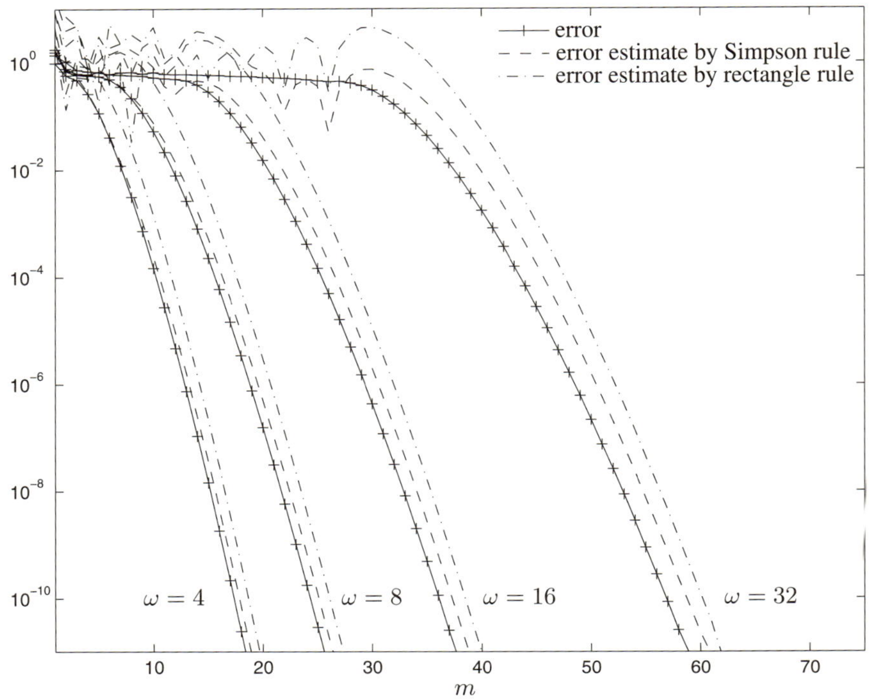

Fig. 2.2. Errors and error estimators (2.21) and (2.22) of the Lanczos method.

Theorem 2.7 (A Posteriori Error Bound). *Let A be a Hermitian matrix, and v a vector of unit Euclidean norm. Then, the error of the mth Lanczos approximation to $e^{-i\Delta t A}v$ is bounded by*

$$\|V_m e^{-i\Delta t T_m} e_1 - e^{-i\Delta t A} v\| \le \tau_{m+1,m} \int_0^{\Delta t} \left| \left[e^{-isT_m} \right]_{m,1} \right| ds\,. \qquad \square$$

If we approximate the integral on the right-hand side by the right-endpoint rectangle rule, we arrive at a *stopping criterion* for the Lanczos iteration (for given Δt) or alternatively at a *step-size selection criterion* (for given m),

$$\Delta t\, \tau_{m+1,m} \left| \left[e^{-i\Delta t\, T_m} \right]_{m,1} \right| \le tol \tag{2.21}$$

for an error tolerance *tol*, or without the factor Δt for an error tolerance per unit step. This criterion has previously been considered with different interpretations by Saad (1992) and Hochbruck, Lubich & Selhofer (1998). In view of Theorem 2.7, a better choice is to take a quadrature rule with more than one function evaluation, for example, the Simpson rule:

$$\Delta t\, \tau_{m+1,m} \left(\frac{2}{3} \left| \left[e^{-i\frac{\Delta t}{2} T_m} \right]_{m,1} \right| + \frac{1}{6} \left| \left[e^{-i\Delta t\, T_m} \right]_{m,1} \right| \right) \le tol. \tag{2.22}$$

With a diagonalized T_m, this is computed at no extra cost.

Example. In Fig. III.2.2 we show the actual errors and the above error estimates of the Lanczos method versus the iteration number m. The method is applied with the tridiagonal matrix $\Delta t A = \frac{\omega}{2}\,\mathrm{tridiag}(-1, 2, -1)$ of dimension 10000 with $\omega = 4, 8, 16, 32$. The eigenvalues of $\Delta t A$ are in the interval $[0, 2\omega]$. The vector v was chosen as a random vector of unit norm. It is instructive to compare the errors with the nearly identical errors in Fig. III.2.1 for the corresponding values of ω. We further note that $\frac{\omega}{2} = c\,\Delta t/\Delta x^2$ can be interpreted as a CFL number for a finite difference discretization of the one-dimensional free Schrödinger equation $i\,\partial_t\psi = -c\,\partial_x^2\psi$.

Lanczos Method for Approximating $f(A)v$. The following lemma follows directly from the Lanczos relations (2.18) and (2.19).

Lemma 2.8. *Let A be a Hermitian matrix and v a vector of unit norm.*
(a) If all eigenvalues of A are in the interval $[a, b]$, then so are those of T_m.
(b) For every polynomial p_{m-1} of degree at most $m - 1$, it holds that

$$p_{m-1}(A)v = V_m\, p_{m-1}(T_m)\, e_1\,. \tag{2.23}$$

Proof. (a) If θ is an eigenvalue of T_m to the eigenvector w of unit norm, then $u = V_m w$ is again of unit norm, and by (2.19), $\theta = w^* T_m w = u^* A u$, which is in $[a, b]$.

(b) Clearly, $v = V_m e_1$. From (2.18) it follows by induction over $k = 1, 2, \ldots$ that

$$A^k V_m e_1 = V_m T_m^k e_1$$

as long as the lower left entry $e_m^T T_m^{k-1} e_1 = 0$. Since T_m^{k-1} is a matrix with $k - 1$ subdiagonals, this holds for $k \le m - 1$. □

For any complex-valued function f defined on $[a, b]$, we have $f(A)$ given via the diagonalization $A = U\,\mathrm{diag}(\lambda_j)U^*$ as $f(A) = U\,\mathrm{diag}(f(\lambda_j))U^*$. Justified by (a) and motivated by (b), we can consider the approximation

$$f(A)v \approx V_m\, f(T_m)\, e_1\,. \tag{2.24}$$

For $f(x) = e^{-i\Delta t\, x}$ this is (2.20). Lemma 2.8 immediately implies the following useful approximation result.

Theorem 2.9 (Optimality of the Lanczos Method). *Let f be a complex-valued function defined on an interval $[a, b]$ that contains the eigenvalues of the Hermitian matrix A, and let v be a vector of unit norm. Then, the error of the Lanczos approximation to $f(A)v$ is bounded by*

$$\|V_m f(T_m) e_1 - f(A)v\| \le 2 \inf_{p_{m-1}} \max_{x\in[a,b]} |p_{m-1}(x) - f(x)|\,,$$

where the infimum is taken over all polynomials of degree at most $m - 1$.

Proof. By Lemma 2.8 (b), we have for every polynomial p_{m-1} of degree at most $m-1$,

$$V_m f(T_m)e_1 - f(A)v = V_m\big(f(T_m) - p_{m-1}(T_m)\big)e_1 - \big(f(A) - p_{m-1}(A)\big)v.$$

Diagonalization of A and T_m and Lemma 2.8 (a) show that each of the two terms to the right is bounded by $\max_{x\in[a,b]} |f(x) - p_{m-1}(x)|$. □

Error Bound of the Lanczos Method for the Matrix Exponential Operator. Combining Theorems 2.9 and 2.2, together with the linear transformation from the interval $[a,b]$ to $[-1,1]$, yields the following result.

Theorem 2.10 (Eventual Superlinear Error Decay). *Let A be a Hermitian matrix all of whose eigenvalues are in the interval $[a,b]$, and let v be a vector of unit Euclidean norm. Then, the error of the Lanczos method (2.20) is bounded by*

$$\|V_m e^{-i\Delta t\, T_m} e_1 - e^{-i\Delta t\, A} v\| \le 8 \left(e^{1-(\omega/2m)^2}\, \frac{\omega}{2m}\right)^m \qquad \text{for} \quad m \ge \omega$$

with $\omega = \Delta t\,(b-a)/2$. □

We remark that the step-size restriction (2.15) of the Chebyshev method applies to the Lanczos method as well.

III.3 Splitting and Composition Methods

The methods of the previous section have the attractive feature that they only require matrix-vector products with the discretized Hamiltonian A of (2.1). However, the maximum permitted step size is inversely proportional to the norm of A, which leads to a time step restriction to $\Delta t = \mathcal{O}(\Delta x^2)$, as we recall from (2.15). The splitting methods considered in this section can achieve good accuracy with no such restriction, provided that the wave function has sufficient spatial regularity.

III.3.1 Splitting Between Kinetic Energy and Potential

We consider the Schrödinger equation

$$i\dot\psi = H\psi \qquad \text{with} \qquad H = T + V\,, \tag{3.1}$$

where T and V are the kinetic energy operator and the potential, respectively, or the corresponding discretized operators. We will assume no bound on the self-adjoint operator or matrix T. In our theoretical results we will assume bounds of the potential V, but the method to be described can work well under weaker assumptions. On the practical side, the basic assumption is that the equations

$$i\dot\theta = T\theta \qquad \text{and} \qquad i\dot\phi = V\phi$$

can both be solved more easily than the full equation (3.1). As we have seen in Chap. I, on the analytical level this is definitely the case in the non-discretized Schrödinger equation: the free Schrödinger equation (only T) is solved by Fourier transformation, and the equation with only the potential V is solved by multiplying the initial data with the scalar exponential $e^{-iV(x)}$ at every space point x. This situation transfers, in particular, to the Fourier collocation method of Section III.1.3, where solving the differential equations for the kinetic and potential parts in (1.32) or (1.33) is done trivially, using the exponentials of diagonal matrices and FFTs.

Strang Splitting. We consider time stepping from an approximation ψ^n at time t^n to the new approximation ψ^{n+1} at time $t^{n+1} = t^n + \Delta t$ by

$$\psi^{n+1} = e^{-i\frac{\Delta t}{2}V} e^{-i\Delta t\, T} e^{-i\frac{\Delta t}{2}V} \psi^n . \tag{3.2}$$

This symmetric operator splitting was apparently first studied by Strang (1968) and independently by Marchuk (1968) in the context of dimensional splitting of advection equations. It was proposed, in conjunction with the Fourier method in space, for nonlinear Schrödinger equations by Hardin & Tappert (1973) and rediscovered for the linear Schrödinger equation, in the disguise of the Fresnel equation of laser optics, by Fleck, Morris & Feit (1976). The scheme was introduced to chemical physics by Feit, Fleck & Steiger (1982). In combination with Fourier collocation in space, the method is usually known as the *split-step Fourier method* in the chemical and physical literature.

Split-Step Fourier Method. In the notation of Sect. III.1.3, we recall the differential equation (1.33) for the vector $u = (u_j)$ of grid values $u_j(t) = \psi_K(x_j, t)$:

$$i\dot{u} = \mathcal{F}_K^{-1} D_K \mathcal{F}_K u + V_K u$$

with the diagonal matrices $D_K = \frac{1}{2\mu}\,\mathrm{diag}(k^2)$ and $V_K = \mathrm{diag}\big(V(x_j)\big)$, where k and j range from $-K/2$ to $K/2 - 1$. With method (3.2), a time step is computed in a way that alternates between pointwise operations and FFTs.

Algorithm 3.1 (Split-Step Fourier Method). *The approximation at time t^n is overwritten by that at time t^{n+1} in the following substeps:*

1. *multiply:* $u_j := e^{-i\frac{\Delta t}{2}V(x_j)} u_j \quad (j = -K/2, \dots, K/2-1)$
2. *FFT:* $u := \mathcal{F}_K u$
3. *multiply:* $u_k := e^{-i\Delta t k^2/(2\mu)} u_k \quad (k = -K/2, \dots, K/2-1)$
4. *inverse FFT:* $u := \mathcal{F}_K^{-1} u$
5. *multiply:* $u_j := e^{-i\frac{\Delta t}{2}V(x_j)} u_j \quad (j = -K/2, \dots, K/2-1)$.

The exponentials in Substep 5 and Substep 1 of the next time step can be combined into a single exponential if the output at time t^{n+1} is not needed.

Unitarity, Symplecticity, Time-Reversibility. The Strang splitting has interesting structure-preserving properties. For self-adjoint T and V, the exponentials $e^{-i\Delta t\, T}$ and $e^{-i\frac{\Delta t}{2}V}$

are unitary (they preserve the norm) and symplectic (they preserve the canonical symplectic two-form $\omega(\xi,\eta) = -2\,\mathrm{Im}\,\langle \xi \,|\, \eta \rangle$, see Theorem II.1.2), and so does their composition. The time-step operator of the Strang splitting is thus both unitary and symplectic. We remark that neither holds for the Chebyshev method, whereas the Lanczos method is unitary, but symplectic only in the restriction to the Krylov subspace, which changes from one time step to the next. Moreover, the Strang splitting is time-reversible: a step of the method starting from ψ^{n+1} with negative step size $-\Delta t$ leads us back to the old ψ_n, or more formally, exchanging $n \leftrightarrow n+1$ and $\Delta t \leftrightarrow -\Delta t$ in the method gives the same method again. We note that neither the Chebyshev method nor the Lanczos method are time-reversible.

III.3.2 Error Bounds for the Strang Splitting

For bounded T and V, Taylor expansion of the exponentials readily shows

$$e^{-i\frac{\Delta t}{2}V}\, e^{-i\Delta t\, T}\, e^{-i\frac{\Delta t}{2}V} = e^{-i\Delta t(T+V)} + \mathcal{O}\big(\Delta t^3 (\|T\| + \|V\|)^3\big)\,.$$

However, such an error bound is of no use when T or V are of large norm. Since $\|T\| \sim (\Delta x)^{-2}$ (as in (2.15)), this error bound would indicate a small error only for $\Delta t \ll \Delta x^2$, whereas numerical experiments clearly indicate that the error of the Strang splitting for initial data of moderately bounded energy is bounded independently of Δx for a given Δt. For problems with smooth potential and smooth initial data the error is numerically observed to be $\mathcal{O}(\Delta t^3)$ uniformly in Δx after one step of the method, and $\mathcal{O}(t^n \Delta t^2)$ at time t^n after n steps, uniformly in n and Δx.

In the following we present an error analysis from Jahnke & Lubich (2000), which explains this favourable behaviour of the splitting method. Here we assume that T and V are self-adjoint operators on a Hilbert space $\mathcal{H}$, and T is positive semi-definite. We require no bound for T, but we assume a (moderate) bound of V:

$$\|V\psi\| \le \beta \|\psi\| \qquad \forall \psi \in \mathcal{H}\,. \tag{3.3}$$

We introduce the norms

$$\begin{aligned} \|\varphi\|_1 &= \langle \varphi \,|\, T + I \,|\, \varphi \rangle^{1/2} \\ \|\varphi\|_2 &= \langle \varphi \,|\, (T+I)^2 \,|\, \varphi \rangle^{1/2} \end{aligned} \tag{3.4}$$

which are the usual Sobolev norms in the case of $T = -\Delta$, and can be viewed as discrete Sobolev norms in the spatially discrete case.

Our main assumptions concern the commutator $[T,V] = TV - VT$ and the repeated commutator $[T,[T,V]]$. We assume that there are constants c_1 and c_2 such that the commutator bounds

$$\|\,[T,V]\varphi\| \;\le\; c_1\, \|\varphi\|_1 \tag{3.5}$$

$$\|\,[T,[T,V]]\varphi\| \;\le\; c_2\, \|\varphi\|_2 \tag{3.6}$$

are satisfied for all φ in a dense domain of $\mathcal{H}$. In the spatially continuous case with $T = -\Delta$ and a potential $V(x)$ that is bounded together with its first- to fourth-order derivatives, we see from the identities

$$\begin{aligned}[\Delta, V]\varphi &= \Delta V\,\varphi + 2\nabla V \cdot \nabla\varphi \\ [\Delta, [\Delta, V]]\varphi &= \Delta^2 V\,\varphi + 4\nabla\Delta V \cdot \nabla\varphi + 4\sum_{j,l} \partial_j\partial_l V\,\partial_j\partial_l\varphi\end{aligned}$$

that the commutator bounds (3.5)–(3.6) are indeed valid. For spatial discretization by the Fourier method, it is shown by Jahnke & Lubich (2000) that these commutator bounds hold with constants c_1 and c_2 that are independent of the discretization parameter. We then have the following second-order error bound.

Theorem 3.2 (Error Bound for the Strang Splitting). *Under the above conditions, the error of the splitting method (3.2) at $t = t^n$ is bounded by*

$$\|\psi^n - \psi(t)\| \le C\,\Delta t^2\, t \max_{0\le\tau\le t} \|\psi(\tau)\|_2\,, \tag{3.7}$$

where C depends only on the bound β of (3.3) and on c_1, c_2 of (3.5)–(3.6).

It is a noteworthy fact that the time discretization error of the splitting method depends on the *spatial* regularity of the wave function, not on its temporal regularity. The proof is done in the usual way by studying the local error of the method (that is, the error after one step) and the error propagation. For the local error we have the following bounds.

Lemma 3.3 (Local Error). *(a) Under conditions (3.3) and (3.5),*

$$\|e^{-i\frac{\Delta t}{2}V}e^{-i\Delta t T}e^{-i\frac{\Delta t}{2}V}\varphi - e^{-i\Delta t(T+V)}\varphi\| \le C_1\,\Delta t^2\,\|\varphi\|_1\,, \tag{3.8}$$

where C_1 depends only on c_1 and β.

(b) Under conditions (3.3) and (3.5)–(3.6),

$$\|e^{-i\frac{\Delta t}{2}V}e^{-i\Delta t T}e^{-i\frac{\Delta t}{2}V}\varphi - e^{-i\Delta t(T+V)}\varphi\| \le C_2\,\Delta t^3\,\|\varphi\|_2\,, \tag{3.9}$$

where C_2 depends only on c_1, c_2 and β.

The local error bound (3.9) together with the telescoping formula

$$\psi^n - \psi(t^n) = S^n\psi^0 - E^n\psi^0 = \sum_{j=0}^{n-1} S^{n-j-1}(S-E)\,E^j\psi^0\,, \tag{3.10}$$

with $S = e^{-i\frac{\Delta t}{2}V}e^{-i\Delta t T}e^{-i\frac{\Delta t}{2}V}$ and $E = e^{-i\Delta t(T+V)}$, immediately yields the error bound of Theorem 3.2. It thus remains to prove the lemma. The basic idea of the following proof is the reduction of the local error to quadrature errors.

Proof. (a) We start from the variation-of-constants formula

$$e^{-i\Delta t(T+V)}\varphi = e^{-i\Delta t T}\varphi - i\int_0^{\Delta t} e^{-isT} V e^{-i(\Delta t - s)(T+V)}\varphi\, ds\,.$$

Expressing the last term under the integral once more by the same formula yields

$$e^{-i\Delta t(T+V)}\varphi = e^{-i\Delta t T}\varphi - i\int_0^{\Delta t} e^{-isT} V e^{-i(\Delta t - s)T}\varphi\, ds + R_1\varphi\,,$$

where the remainder

$$R_1 = -\int_0^{\Delta t} e^{sT} V \int_0^{\Delta t - s} e^{-i\sigma T} V e^{-i(\Delta t - s - \sigma)(T+V)}\, d\sigma\, ds$$

is bounded in the operator norm by $\|R_1\| \le \frac{1}{2}\Delta t^2\beta^2$. On the other hand, using the exponential series for $e^{-i\frac{\Delta t}{2}V}$ leads to

$$e^{-i\frac{\Delta t}{2}V} e^{-i\Delta t T} e^{-i\frac{\Delta t}{2}V}\varphi = e^{-i\Delta t T}\varphi - \frac{i}{2}\Delta t\big(V e^{-i\Delta t T} + e^{-i\Delta t T} V\big)\varphi + R_2\varphi\,,$$

where $\|R_2\| \le \frac{1}{2}\Delta t^2\beta^2$. The basic observation is now that the second term is the trapezoidal rule approximation to the integral appearing for the exact solution, with the integrand $f(s) = -i\, e^{-isT} V e^{-i(\Delta t - s)T}\varphi$. Consequently, the error is of the form

$$e^{-i\frac{\Delta t}{2}V} e^{-i\Delta t T} e^{-i\frac{\Delta t}{2}V}\varphi - e^{-i\Delta t(T+V)}\varphi = d + r\,, \tag{3.11}$$

where $r = R_2\varphi - R_1\varphi$ collects the remainder terms and

$$\begin{aligned} d &= \frac{1}{2}\Delta t\,\big(f(0) + f(\Delta t)\big) - \int_0^{\Delta t} f(s)\, ds \\ &= -\Delta t^2 \int_0^1 \big(\tfrac{1}{2} - \theta\big)\, f'(\theta\Delta t)\, d\theta = \frac{1}{2}\,\Delta t^3 \int_0^1 \theta(1-\theta) f''(\theta\Delta t)\, d\theta \end{aligned} \tag{3.12}$$

is the error of the trapezoidal rule, written in first- and second-order Peano form. Since $f'(s) = -e^{-isT}[T,V]e^{-i(\Delta t - s)T}\varphi$, condition (3.5) yields the error bound (3.8).

(b) For the error bound (3.9), we use $f''(s) = i\, e^{-isT}[T,[T,V]]e^{-i(\Delta t - s)T}\varphi$ and condition (3.6) to bound

$$\|d\| \le \frac{1}{12}\, c_2\, \Delta t^3\, \|\varphi\|_2\,. \tag{3.13}$$

It remains to study $r = R_2 v - R_1 v$. We have

$$R_1 = -\int_0^{\Delta t} e^{-isT} V \int_0^{\Delta t - s} e^{-i\sigma T} V e^{-i(\Delta t - s - \sigma)T}\, d\sigma\, ds + \widetilde{R}_1$$

with $\|\widetilde{R}_1\| \le C\Delta t^3\beta^3$, and

$$R_2 = -\frac{1}{8}\,\Delta t^2\,\left(V^2 e^{-i\Delta t T} + 2V e^{-i\Delta t T} V + e^{-i\Delta t T} V^2\right) + \widetilde{R}_2$$

with $\|\widetilde{R}_2\| \le C\Delta t^3 \beta^3$. We thus obtain

$$r = \widetilde{d} + \widetilde{r}\,, \tag{3.14}$$

where $\widetilde{r} = \widetilde{R}_2\varphi - \widetilde{R}_1\varphi$ is bounded by $\|\widetilde{r}\| \le C\Delta t^3\,\beta^3\,\|\varphi\|$ and, with $g(s,\sigma) = -e^{-isT} V e^{-i\sigma T} V e^{-i(\Delta t - s - \sigma)T}\varphi$,

$$\widetilde{d} = \frac{1}{8}\,\Delta t^2\,\Big(g(0,0) + 2g(0,\Delta t) + g(\Delta t, 0)\Big) - \int_0^{\Delta t}\int_0^{\Delta t - s} g(s,\sigma)\,d\sigma\,ds$$

is the error of a quadrature formula that integrates constant functions exactly. Hence,

$$\|\widetilde{d}\| \le \widetilde{c}\,\Delta t^3 \left(\max\left\|\frac{\partial g}{\partial s}\right\| + \max\left\|\frac{\partial g}{\partial \sigma}\right\|\right),$$

where the maxima are taken over the triangle $0 \le s \le \Delta t$, $0 \le \sigma \le \Delta t - s$. Since

$$\frac{\partial g}{\partial s}(s,\sigma) = i\,e^{-isT}[T,V]e^{-i\sigma T} V e^{-i(\Delta t - s - \sigma)T}\varphi + i\,e^{-isT} V e^{-i\sigma T}[T,V]e^{-i(\Delta t - s - \sigma)T}\varphi\,,$$

we obtain, using (3.5),

$$\left\|\frac{\partial g}{\partial s}\right\| \le c_1\,(c_1 + \beta)\,\|\varphi\|_1 + \beta\,c_1\,\|\varphi\|_1\,.$$

Similarly, $\|\partial g/\partial\sigma\| \le \beta\,c_1\,\|\varphi\|_1$, so that finally

$$\|\widetilde{d}\| \le C\Delta t^3\,\|\varphi\|_1\,.$$

Together with the above bounds for $\widetilde{r}$ and d this yields the error bound (3.9). □

An error analysis of splitting methods in the situation of two unbounded operators, and in particular, including cases of the Schrödinger equation with an unbounded potential, is given by Hansen & Ostermann (2008). Their error bounds show the full temporal order of approximation but require higher regularity than Theorem 3.2.

III.3.3 Higher-Order Compositions

The Strang splitting $S(\Delta t) = e^{-i\frac{\Delta t}{2}V} e^{-i\Delta t T} e^{-i\frac{\Delta t}{2}V}$ yields a second-order method. Higher-order methods can be obtained by a suitable composition of steps of different size of the basic method:

$$\psi^{n+1} = S(\gamma_s \Delta t)\ldots S(\gamma_1 \Delta t)\psi^n \tag{3.15}$$

with symmetrically arranged coefficients $\gamma_j = \gamma_{s+1-j}$ determined such that

$$S(\gamma_s \Delta t) \dots S(\gamma_1 \Delta t) = e^{-i\Delta t(T+V)} + \mathcal{O}\big(\Delta t^{p+1}(\|T\| + \|V\|)^{p+1}\big)$$

with an order $p > 2$. Composition methods of this or similar type have been devised by Suzuki (1990) and Yoshida (1990), and improved methods have since been constructed, e.g., by McLachlan (1995), Kahan & Li (1997), Blanes & Moan (2002), Sofroniou & Spaletta (2005). We refer to Hairer, Lubich & Wanner (2006), Sect. V.3, and McLachlan & Quispel (2002) for reviews of composition methods, for their order theory, for their coefficients, and for further references. For example, an excellent method of order $p = 8$ with $s = 17$ by Kahan & Li (1997) has the coefficients

$$\begin{aligned}
\gamma_1 = \gamma_{17} &= 0.13020248308889008087881763 \\
\gamma_2 = \gamma_{16} &= 0.56116298177510838456196441 \\
\gamma_3 = \gamma_{15} &= -0.38947496264484728640807860 \\
\gamma_4 = \gamma_{14} &= 0.15884190655515560089621075 \\
\gamma_5 = \gamma_{13} &= -0.39590389413323757733623154 \\
\gamma_6 = \gamma_{12} &= 0.18453964097831570709183254 \\
\gamma_7 = \gamma_{11} &= 0.25837438768632204729397911 \\
\gamma_8 = \gamma_{10} &= 0.29501172360931029887096624 \\
\gamma_9 &= -0.60550853383003451169892108
\end{aligned} \tag{3.16}$$

As with the basic Strang splitting method, the presence of powers of $\|T\|$ in the error bound would seem to make a step-size restriction $\Delta t \ll \Delta x^2$ necessary, but indeed this is not the case. Thalhammer (**?**) proves high-order error bounds for such methods that require no bound of T. By a formidable extension of the approach in the proof of Theorem 3.2, using p-fold repeated-commutator bounds and achieving a reduction to quadrature errors, it is shown that in the spatially continuous case with $T = -\Delta$ and a smooth bounded potential, there is a pth-order error bound at $t = t^n$

$$\|\psi^n - \psi(t)\| \le C\, \Delta t^p\, t \max_{0\le\tau\le t} \|\psi(\tau)\|_p \tag{3.17}$$

with the pth-order Sobolev norm. It is to be expected that in the spatially discretized case, the required commutator bounds hold uniformly in Δx so that the error bound becomes uniform in the spatial discretization parameter.

III.4 Integrators for Time-Dependent Hamiltonians

In contrast to the Chebyshev and Lanczos methods, splitting methods extend directly to the Schrödinger equation (0.1) with a time-dependent potential $V(x,t)$.

Strang Splitting. For a time-dependent potential $V(t) = V(\cdot, t)$, a version of this method reads

$$\psi^{n+1} = e^{-i\frac{\Delta t}{2}\, V(t^{n+1})}\, e^{-i\Delta t\, T}\, e^{-i\frac{\Delta t}{2}\, V(t^n)}\, \psi^n\,. \tag{4.1}$$

The error analysis of Theorem 3.2 is straightforwardly extended to the case of a time-dependent bounded, smooth potential, where one still obtains a second-order error bound (3.7), viz.,

$$\|\psi^n - \psi(t)\| \le C\,\Delta t^2\, t \max_{0\le\tau\le t} \|\psi(\tau)\|_2 .$$

Higher-Order Compositions. If we denote $S(t, \Delta t) = e^{-i\frac{\Delta t}{2}V(t+\Delta t)}\, e^{-i\Delta t\, T}\, e^{-i\frac{\Delta t}{2}V(t)}$, then the composition method

$$\psi^{n+1} = S(t^n + \theta_s \Delta t, \gamma_s \Delta t) \dots S(t^n + \theta_1 \Delta t, \gamma_1 \Delta t)\psi^n$$

with $\theta_1 = 0$ and $\theta_{k+1} = \theta_k + \gamma_k$ (for $k = 1, \dots, s-1$) has the same formal order p as method (3.15) for all the methods proposed in the papers cited above. It is to be expected that the error analysis by Thalhammer (**?**) can be extended to yield the full-order error bound (3.17) also in the case of smooth time-dependent potentials.

Magnus Methods. This approach, which has its origin in work by Magnus (1954), uses an approximation to the solution of

$$i\dot{\psi}(t) = H(t)\psi(t)$$

by an exponential

$$\psi^{n+1} = e^{\Omega^n}\,\psi^n , \tag{4.2}$$

where suitable choices for Ω^n are, for example, the mid-point rule

$$\Omega^n = -i\Delta t\, H\bigl(t^n + \frac{\Delta t}{2}\bigr) \tag{4.3}$$

or the method based on the two-stage Gaussian quadrature with nodes $c_{1,2} = \frac{1}{2} \pm \frac{\sqrt{3}}{6}$,

$$\Omega^n = -\frac{i}{2}\,\Delta t\,\bigl(H_1 + H_2\bigr) - \frac{\sqrt{3}}{12}\,\Delta t^2\,\bigl[H_2, H_1\bigr] \tag{4.4}$$

with $H_j = H(t^n + c_j \Delta t)$ for $j = 1, 2$.

We refer to Iserles & Nørsett (1999) and Iserles, Munthe-Kaas, Nørsett & Zanna (2000) for the theory of Magnus-type methods for *bounded* $H(t)$ (more precisely, for $\Delta t\,\|H(t)\| \to 0$), and to Blanes, Casas & Ros (2000) for the construction of efficient high-order Magnus methods. Various interesting commutator-free fourth-order methods for time-dependent Hamiltonians are given by Blanes & Moan (2000).

For the Schrödinger equation with $H(t) = T + V(t)$ for a (discretized) negative Laplacian T and a smooth time-dependent potential, it is shown by Hochbruck & Lubich (2003) that the Magnus methods retain their full order of convergence (without bounds of T entering the error bound) if the solution is sufficiently regular. The error analysis again uses commutator bounds similar to (3.5) and (3.6). In particular, the methods (4.3) and (4.4) are of temporal orders 2 and 4, respectively, uniformly with respect to the space discretization. The error bounds at $t = t^n$ are

$$\|\psi^n - \psi(t)\| \le C\,\Delta t^2\, t \max_{0\le\tau\le t} \|\psi(\tau)\|_1$$

for method (4.3), and

$$\|\psi^n - \psi(t)\| \le C\,\Delta t^4\, t \max_{0\le\tau\le t} \|\psi(\tau)\|_k$$

with $k = 7$ for method (4.4). In the spatially discretized case this improves to $k = 3$ if $\Delta t\,\|T^{1/2}\| \le C$, which amounts to a mild step size restriction $\Delta t = \mathcal{O}(\Delta x)$.

In a practical implementation, the matrix exponential times a vector in (4.2) is approximated by the Chebyshev or Lanczos method, which only requires the action of Ω^n on vectors, or once again by splitting methods.

Chapter IV.
Numerical Methods for Non-Linear Reduced Models

In this chapter we turn to numerical methods for non-linear reduced models that result from a variational approximation as considered in Chapter II. We first study the space discretization within the variational framework, which amounts to a further reduction of the approximation manifold in the Dirac–Frenkel time-dependent variational principle to a finite-dimensional manifold defined in terms of fixed basis functions. For time discretization, we discuss a splitting approach that applies directly to the formulation by the Dirac–Frenkel variational principle rather than the equations of motion of the particular model. We first give an abstract formulation of the variational splitting method and then apply it to multi-configuration time-dependent Hartree (MCTDH) and Gaussian wave-packet dynamics.

The general approach in this section is to stay within the variational framework as far as possible, and to commit variational crimes only at the final stages, e.g., in actually computing integrals for the matrix elements and in special time-stepping methods, where the effect of the non-variational perturbations can be numerically controlled.

IV.1 Variational Space Discretization

Variational approximation methods such as the time-dependent Hartree method or its multi-configuration version leave us with nonlinear partial differential equations, which still need to be discretized in space and time. Rather than choosing some *ad hoc* space discretization of the equations of motion, we here consider using once again the Dirac–Frenkel variational approximation principle to arrive at the spatially discretized equations.

IV.1.1 Abstract Formulation

We return to the abstract setting of Section II.1 and consider a Schrödinger equation on a complex Hilbert space $\mathcal{H}$ with inner product $\langle\cdot|\cdot\rangle$, with a Hamiltonian H that is a self-adjoint linear operator on $\mathcal{H}$,

$$\frac{d\psi}{dt} = \frac{1}{i\hbar} H\psi \,. \tag{1.1}$$

Let $\mathcal{M}$, a submanifold of $\mathcal{H}$, be the approximation manifold on which an approximate solution $u(t)$ to the solution $\psi(t)$ of (1.1) with initial data $u(0) = \psi(0) \in \mathcal{M}$ is sought.

The time derivative to the approximate wave function $u(t) \in \mathcal{M}$ is determined from the Dirac–Frenkel variational principle (II.1.2), viz.,

$$\frac{du}{dt} \in \mathcal{T}_u\mathcal{M} \quad \text{such that} \quad \Big\langle v \,\Big|\, \frac{du}{dt} - \frac{1}{i\hbar} H u \Big\rangle = 0 \quad \forall\, v \in \mathcal{T}_u\mathcal{M}\,, \tag{1.2}$$

where $\mathcal{T}_u\mathcal{M}$ is the tangent space at $u \in \mathcal{M}$, henceforth assumed to be complex linear and to have $u \in \mathcal{T}_u\mathcal{M}$.

We now consider a family of finite-dimensional manifolds $\mathcal{M}_K \subset \mathcal{M}$ with a discretization parameter K, which approximate $\mathcal{M}$ as $K \to \infty$:

$$\text{For every } u \in \mathcal{M},\ \inf_{w_K \in \mathcal{M}_K} \|w_K - u\| \to 0 \text{ as } K \to \infty. \tag{1.3}$$

We then discretize (1.2) by using the variational approximation on $\mathcal{M}_K$: find an approximate wave function $u_K(t) \in \mathcal{M}_K$ with

$$\frac{du_K}{dt} \in \mathcal{T}_{u_K}\mathcal{M}_K \quad \text{such that} \quad \Big\langle v_K \,\Big|\, \frac{du_K}{dt} - \frac{1}{i\hbar} H u_K \Big\rangle = 0 \quad \forall\, v_K \in \mathcal{T}_{u_K}\mathcal{M}_K. \tag{1.4}$$

We exemplify this procedure in the Hartree model and then return to study the error $u_K(t) - u(t)$ in the abstract setting.

IV.1.2 Space Discretization of the Hartree and MCTDH Equations

Time-Dependent Hartree Approximation. We recall from Sect. II.3.1 that in the Hartree method the approximations are chosen as Hartree products of single-particle functions, which lie in

$$\mathcal{M} = \big\{ u \in L^2(\mathbb{R}^{dN}) \,:\, u \neq 0,\ u = a\,\varphi_1 \otimes \cdots \otimes \varphi_N\,,\ a \in \mathbb{C},\ \varphi_n \in L^2(\mathbb{R}^d) \big\}. \tag{1.5}$$

We now approximate each single-particle function φ_n by a finite linear combination of basis functions $\chi_k^{(n)} \in L^2(\mathbb{R}^d)$, which in the following we assume orthonormal for ease of presentation, for example, tensor products of Hermite functions as in Sect. III.1. We denote

$$\mathcal{V}_K^{(n)} = \mathrm{span}\big(\chi_k^{(n)} : k = 1, \dots, K\big)\,,$$

where for simplicity we choose K independent of n. We thus have the finite-dimensional approximation manifold

$$\mathcal{M}_K = \big\{ u_K \in L^2(\mathbb{R}^{3N}) \,:\, u_K \neq 0,\ u_K = a\,\varphi_K^{(1)} \otimes \cdots \otimes \varphi_K^{(N)}\,,\ a \in \mathbb{C},\ \varphi_K^{(n)} \in \mathcal{V}_K^{(n)} \big\}. \tag{1.6}$$

Retracing the derivation of the Hartree equations of motion down to (II.3.9), we now arrive at

$$\begin{aligned} \Big\langle \vartheta_K^{(n)} \,\Big|\, \frac{\partial \varphi_K^{(n)}}{\partial t} \Big\rangle &= \Big\langle a\,\varphi_K^{(1)} \otimes \cdots \otimes \vartheta_K^{(n)} \otimes \cdots \otimes \varphi_K^{(N)} \,\Big|\, \frac{1}{i\hbar} H u_K \Big\rangle \\ &\quad - \Big\langle u_K \,\Big|\, \frac{1}{i\hbar} H u_K \Big\rangle \langle \vartheta_K^{(n)} \,|\, \varphi_K^{(n)} \rangle \qquad \forall\, \vartheta_K^{(n)} \in \mathcal{V}_K^{(n)}\,, \end{aligned} \tag{1.7}$$

where simply the space L^2 has been replaced by the finite-dimensional approximation space $\mathcal{V}_K^{(n)}$. Writing

$$\varphi_K^{(n)}(x_n,t) = \sum_{k=1}^{K} c_k^{(n)}(t)\,\chi_k^{(n)}(x_n)\,,$$

this is readily shown to yield the following system of ordinary differential equations for the coefficients (as in (II.3.12), we ignore a phase term):

$$i\hbar\,\frac{dc_k^{(n)}}{dt} = \sum_{l=1}^{K} \langle \chi_k^{(n)} \,|\, H^{(n)} \,|\, \chi_l^{(n)} \rangle^{(n)}\, c_l^{(n)} \tag{1.8}$$

with the mean-field Hamiltonian

$$H^{(n)} = \langle \psi_K^{(n)} \,|\, H \,|\, \psi_K^{(n)} \rangle^{(\neg n)} \qquad \text{with} \quad \psi_K^{(n)} = \bigotimes_{j\neq n} \varphi_K^{(j)}\,,$$

where the superscript $(\neg n)$ indicates that the L^2 inner product on the right-hand side is over all variables except x_n. We note that equations (1.8) are just a Galerkin discretization of the Hartree equations (II.3.12).

MCTDH Approximation. Conceptually the same is done for the MCTDH method. Dropping the discretization parameter K in the notation, we approximate by a linear combination of Hartree products,

$$u(x_1,\ldots,x_N,t) \approx \sum_J a_J(t)\,\varphi_{j_1}^{(1)}(x_1,t)\cdot\ldots\cdot\varphi_{j_N}^{(N)}(x_N,t)$$

with the sum over multi-indices $J=(j_1,\ldots,j_N)$, where now each single-particle function $\varphi_{j_n}^{(n)}$ is a finite linear combination of basis functions:

$$\varphi_j^{(n)}(x_n,t) = \sum_{\alpha=1}^{K} c_{j,\alpha}^{(n)}(t)\,\chi_\alpha^{(n)}(x_n)\,.$$

By the arguments of Sect. II.3.3, we find that the coefficients satisfy differential equations which, in the notation of Theorem II.3.4, read

$$\begin{aligned}
i\hbar\,\frac{da_J}{dt} &= \sum_I \langle \Phi_J | H | \Phi_I \rangle\, a_I\,, \qquad\qquad J=(j_1,\ldots,j_N)\,,\\
i\hbar\,\frac{dc_{j,\alpha}^{(n)}}{\partial t} &= \sum_{k=1}^{r_n}\sum_{l=1}^{r_n}\sum_{\beta=1}^{K} (\rho^{(n)})^{-1}_{j,k} \langle \chi_\alpha^{(n)} \,|\, H_{kl}^{(n)} \,|\, \chi_\beta^{(n)} \rangle^{(n)}\, c_{l,\beta}^{(n)}\\
&\quad - \sum_{k=1}^{r_n}\sum_{l=1}^{r_n}\sum_{m=1}^{r_n}\sum_{\beta=1}^{K}\sum_{\gamma=1}^{K} (\rho^{(n)})^{-1}_{j,k}\, c_{m,\alpha}^{(n)}\, \overline{c_{m,\gamma}^{(n)}}\, \langle \chi_\gamma^{(n)} \,|\, H_{kl}^{(n)} \,|\, \chi_\beta^{(n)} \rangle^{(n)}\, c_{l,\beta}^{(n)}\\
&\qquad\qquad j=1,\ldots,r_n,\; n=1,\ldots,N\,,\; \alpha=1,\ldots,K\,,
\end{aligned}$$

where we have the mean-field operators $H_{kl}^{(n)} = \langle \psi_k^{(n)} \,|\, H \,|\, \psi_l^{(n)} \rangle^{(\neg n)}$ with the single-hole functions $\psi_j^{(n)}$ of (II.3.33).

IV.1.3 Discretization Error

We study the error $\|u_K(t) - u(t)\|$ under the assumptions of Sect. II.6 for the manifolds $\mathcal{M}_K$, with constants that are uniform in K. These assumptions can be verified, e.g., for variational Hermite or Fourier discretizations of the Hartree and MCTDH models. We let again $\hbar = 1$ in this subsection. The Hamiltonian H is split as

$$H = A + B \tag{1.9}$$

with self-adjoint linear operators A and B where B is bounded,

$$\|B\varphi\| \leq \beta \,\|\varphi\| \qquad \text{for all} \quad \varphi \in \mathcal{H}\,, \tag{1.10}$$

and A is tangential on $\mathcal{M}_K$:

$$Au_K \in \mathcal{T}_{u_K}\mathcal{M}_K \qquad \text{for all} \quad u_K \in \mathcal{M}_K. \tag{1.11}$$

The orthogonal projectors $P_K(u_K)$ onto $\mathcal{T}_{u_K}\mathcal{M}_K$ and $P_K^\perp(u_K) = I - P_K(u_K)$ satisfy

$$\begin{aligned} \|\,(P_K(u_K) - P_K(v_K))\varphi\,\| &\leq \kappa\,\|u_K - v_K\| \cdot \|\varphi\| && (1.12)\\ \|\,P_K^\perp(v_K)(u_K - v_K)\,\| &\leq \kappa\,\|u_K - v_K\|^2 && (1.13) \end{aligned}$$

for all $u_K, v_K \in \mathcal{M}_K$ and $\varphi \in \mathcal{H}$. We further assume

$$\text{dist}(u(t), \mathcal{M}_K) \leq \frac{1}{2\kappa} \qquad \text{for} \quad 0 \leq t \leq \bar{t}\,, \tag{1.14}$$

which is satisfied for sufficiently large K under the approximation condition (1.3). We require the bounds

$$\|Hu(t)\| \leq \mu\,, \qquad \|Hu_K(t)\| \leq \mu \quad \text{and} \quad \|Au_K(t)\| \leq \mu\,. \tag{1.15}$$

Further we consider the distance bound $\delta \leq \mu$ given by

$$\text{dist}\,(Hu(t), \mathcal{T}_{v_K(t)}\mathcal{M}) \leq \delta\,, \qquad \text{dist}\,(Hu_K(t), \mathcal{T}_{u_K(t)}\mathcal{M}) \leq \delta\,, \tag{1.16}$$

where $v_K(t) \in \mathcal{M}_K$ is the best approximation to $u(t)$ on $\mathcal{M}_K$:

$$d_K(t) = \text{dist}\,(u(t), \mathcal{M}_K) = \|v_K(t) - u(t)\|\,.$$

We then have the following extension of Theorem II.6.1, where we note in addition $\|u_K(t) - u(t)\| \leq \|u_K(t) - v_K(t)\| + d_K(t)$.

Theorem 1.1 (Quasi-Optimality). *Under conditions (1.9)–(1.16), the difference between the variational discretization $u_K(t)$ of (1.4) and the best approximation $v_K(t)$ on $\mathcal{M}_K$ is bounded by*

$$\|u_K(t) - v_K(t)\| \leq e^{\gamma t}\|u_K(0) - v_K(0)\| + Ce^{\gamma t}\int_0^t d_K(s)\,ds \tag{1.17}$$

with $\gamma = 2\kappa\delta$ and $C = \beta + 4\kappa\mu$, for $0 \leq t \leq \bar{t}$.

Proof. The proof is almost identical to that of Theorem II.6.1, where now u_K and u assume the roles of u and ψ, respectively. We compare the differential equation for u_K,

$$\dot{u}_K = P_K(u_K)\frac{1}{i} H u_K \,,$$

with that for the best approximation v_K, which in correspondence with (II.6.14) reads

$$\dot{v}_K = P_K(v_K)\dot{u} + r(v_K, u) \qquad \text{with} \quad \|r(v_K, u)\| \le 2\kappa\mu d_K \,.$$

Since $\dot{u} = P(u)\frac{1}{i} Hu$, and $P_K(v_K)P(v_K) = P_K(v_K)$ because of $\mathcal{M}_K \subset \mathcal{M}$, we obtain

$$\dot{v}_K = P_K(v_K)\frac{1}{i} H v_K - P_K(v_K)\frac{1}{i} H(v_K - u) + P_K(v_K)(P(u) - P(v_K))\frac{1}{i} Hu \,,$$

where the last term is bounded by $\kappa\mu d_K$. The proof then proceeds in the same way as that of Theorem II.6.1. □

For the Hartree or MCTDH model on a bounded interval in each coordinate direction and with periodic boundary conditions, Theorem 1.1 can be used to show that the variational Fourier discretization with K Fourier modes in each coordinate satisfies an error bound

$$\|u_K(t) - u(t)\| \le C(t)\, K^{-s} \max_{0 \le \tau \le t} \|u(\tau)\|_{H^s}$$

if the Hartree or MCTDH wave function u is in the periodic Sobolev space of order s. It is to be expected that this estimate can be extended to the Fourier collocation method via the interpretation of collocation as a Galerkin method with quadrature approximation of the matrix elements.

IV.2 Variational Splitting: Abstract Formulation

Splitting methods were found to be useful as time integration methods for the linear Schrödinger equation (Sect. III.3). Here, they are extended to variational approximations.

IV.2.1 Splitting the Variational Equation

Suppose that

$$H = T + V \tag{2.1}$$

and that the variational equations (1.2) with T and V instead of H are easier to solve than that for the full Hamiltonian H. Then, the following splitting approach appears promising.

Algorithm 2.1 (Variational Splitting). *A time step from $u^n \in \mathcal{M}$ to the new approximation $u^{n+1} \in \mathcal{M}$ at time $t^{n+1} = t^n + \Delta t$ is done as follows.*

1. Half-step with V: *Determine $u_+^n \in \mathcal{M}$ as the solution at time $\Delta t/2$ of the equation for u,*
$$\frac{du}{dt} \in \mathcal{T}_u\mathcal{M} \quad \textit{such that} \quad \Big\langle v \,\Big|\, \frac{du}{dt} - \frac{1}{i\hbar}Vu \Big\rangle = 0 \quad \forall\, v \in \mathcal{T}_u\mathcal{M}\,, \tag{2.2}$$
with initial value $u(0) = u^n \in \mathcal{M}$.
2. Full step with T: *Determine u_-^{n+1} as the solution at time Δt of*
$$\frac{du}{dt} \in \mathcal{T}_u\mathcal{M} \quad \textit{such that} \quad \Big\langle v \,\Big|\, \frac{du}{dt} - \frac{1}{i\hbar}Tu \Big\rangle = 0 \quad \forall\, v \in \mathcal{T}_u\mathcal{M}\,, \tag{2.3}$$
with initial value $u(0) = u_+^n$.
3. Half-step with V : *Finally, u^{n+1} is the solution at time $\Delta t/2$ of (2.2) with initial value $u(0) = u_-^{n+1}$.*

This method was put forward in Lubich (2004) as a numerical integrator for the MCTDH approximation. We note that for $\mathcal{M} = \mathcal{H}$ this is just the Strang splitting (III.3.2). When we compose steps of different length by this method as in Section III.3.3, we again obtain higher-order methods (at least formally).

Norm Preservation, Symplecticity, Time Reversibility. The above method is a symmetric composition of exact flows that preserve norm and symplecticity according to Section II.1. Disregarding additional numerical errors that may result from an inexact solution of the differential equations (2.2) and (2.3), the map $u^n \mapsto u^{n+1}$ is therefore norm-preserving, symplectic, and time-reversible.

IV.2.2 Error Analysis

We give a non-linear extension of the error analysis of Theorem III.3.2 to derive a second-order error bound. In this subsection we let $\hbar = 1$ for convenience.

Assumptions. Our assumptions are similar to those of Sect. II.6. We consider the situation that the approximation manifold $\mathcal{M}$ is such that solutions of the free Schrödinger equation starting on $\mathcal{M}$ remain on $\mathcal{M}$:
$$e^{-itT}u \in \mathcal{M} \qquad \text{for} \quad u \in \mathcal{M},\ t \in \mathbb{R}\,.$$
This is satisfied for all the reduced models of Chapter II. An equivalent condition is
$$Tu \in \mathcal{T}_u\mathcal{M} \qquad \text{for} \quad u \in \mathcal{M} \cap D(T)\,. \tag{2.4}$$
We assume a bound on V:
$$\|V\varphi\| \le \beta\,\|\varphi\| \qquad \forall \varphi \in \mathcal{H}\,. \tag{2.5}$$

We assume that the manifold has bounded curvature: the orthogonal projections $P(u) : \mathcal{H} \to \mathcal{T}_u\mathcal{M}$ and $P^\perp(u) = I - P(u)$ satisfy (II.6.5)–(II.6.6), viz.,

$$\| (P(u) - P(v))\varphi \| \leq \kappa \|u - v\| \cdot \|\varphi\| \tag{2.6}$$

$$\| P^\perp(v)(u - v) \| \leq \kappa \|u - v\|^2 \tag{2.7}$$

for all $u, v \in \mathcal{M}$ and $\varphi \in \mathcal{H}$. In addition we need non-linear versions of the commutator bounds (III.3.5)–(III.3.6). For their formulation we need to introduce some more notation.

Vector Fields and Flows on $\mathcal{M}$. The exact solution of (1.2) is

$$u(t) = \phi_H^t(u^0)$$

where ϕ_H^t is the *flow* map of the differential equation on $\mathcal{M}$,

$$\dot{u} = \widehat{H}(u) \qquad \text{with} \quad \widehat{H}(u) = P(u)\frac{1}{i}Hu\,.$$

that is, $\phi_H^t(v) \in \mathcal{M}$ is the solution at time t of this differential equation with initial value $u(0) = v \in \mathcal{M}$. A step of the splitting algorithm reads, in similar notation,

$$u^{n+1} = \phi_V^{\Delta t/2} \circ \phi_T^{\Delta t} \circ \phi_V^{\Delta t/2}(u^n)\,, \tag{2.8}$$

where ϕ_V^t is the flow on $\mathcal{M}$ of

$$\dot{u} = \widehat{V}(u) \qquad \text{with} \quad \widehat{V}(u) = P(u)\frac{1}{i}Vu,$$

and $\phi_T^t = e^{-itT}$ is the flow on $\mathcal{M}$ of

$$\dot{u} = \widehat{T}(u) \qquad \text{with} \quad \widehat{T}(u) = P(u)\frac{1}{i}Tu = \frac{1}{i}Tu\,.$$

The last equality holds by condition (2.4).

Lie-Commutator Bounds. We require bounds on the Lie commutator of the vector fields $\widehat{T}$ and $\widehat{V}$ on $\mathcal{M}$, given by

$$[\widehat{T}, \widehat{V}](u) = \frac{1}{i}T\widehat{V}(u) - \widehat{V}'(u)\frac{1}{i}Tu$$

with the directional derivative

$$\widehat{V}'(u)\frac{1}{i}Tu = \frac{d}{dt}\Big|_{t=0} \widehat{V}(e^{tT/i}u).$$

A further estimate is needed for the iterated commutator $[\widehat{T}, [\widehat{T}, \widehat{V}]]$. We assume non-linear versions of the bounds (III.3.5)–(III.3.6):

$$\|[\widehat{T}, \widehat{V}](u)\| \leq c_1\|u\|_1 \tag{2.9}$$

$$\|[\widehat{T}, [\widehat{T}, \widehat{V}]](u)\| \leq c_2\|u\|_2 \tag{2.10}$$

with constants c_1, c_2 independent of $u \in \mathcal{M} \cap D(T)$. Here, the Sobolev-type norms with subscripts 1 and 2 are again those defined by (III.3.4). These conditions indeed turn out to be satisfied for the Hartree method and its multi-configuration versions in the case of a smooth bounded potential; see Lubich (2004).

Error Bound for the Variational Splitting. Following Lubich (2004), we obtain a second-order error bound of the type of Theorem III.3.2.

Theorem 2.2 (Error Bound). *Under conditions (2.4)–(2.10), the error of the variational splitting method is bounded at* $t = t^n$ *by*

$$\|u^n - u(t)\| \le C(t)\, \Delta t^2 \max_{0\le\tau\le t} \|u(\tau)\|_2\,,$$

where $C(t)$ *depends only on* β, c_1, c_2, κ, *and* t.

The proof is done in the usual pattern by combining estimates of the local error and a stability estimate.

Lemma 2.3 (Local Error). *In the situation of Theorem 2.2, the error after one step is bounded by*

$$\|u^1 - u(\Delta t)\| \le C_1\, \Delta t^2 \max_{0\le\tau\le \Delta t} \|u(\tau)\|_1 \tag{2.11}$$

$$\|u^1 - u(\Delta t)\| \le C_2\, \Delta t^3 \max_{0\le\tau\le \Delta t} \|u(\tau)\|_2\,, \tag{2.12}$$

where C_1 *and* C_2 *depend only on* β, c_1, c_2.

Lemma 2.4 (Stability). *Let* u^1 *and* v^1 *be the numerical solutions after one step starting from* $u^0 \in \mathcal{M}$ *and* $v^0 \in \mathcal{M}$, *resp., with* u^0 *and* v^0 *of unit norm and* $\|u^0 - v^0\| \le c\Delta t$. *Then, their difference is bounded by*

$$\|u^1 - v^1\| \le e^{\gamma \Delta t}\, \|u^0 - v^0\| \tag{2.13}$$

with $\gamma = \kappa\delta + \mathcal{O}(\Delta t)$, *where* $\delta = \mathrm{dist}(Vv^0, \mathcal{T}_{v^0}\mathcal{M})$.

If we denote a step of the variational splitting method by $u^{n+1} = S_{\Delta t}(u^n)$, then the error accumulation formula (Lady Windermere's Fan; see Hairer, Nørsett & Wanner (1993), Sect. II.3)

$$u^n - u(t^n) = \sum_{j=0}^{n-1} \Big(S_{\Delta t}^{n-1-j}\big(S_{\Delta t}(u(t^j)\big) - S_{\Delta t}^{n-1-j}\big((u(t^{j+1})\big)\Big)$$

together with Lemmas 2.3 and 2.4 yield Theorem 2.2 with $C(t) = (e^{\gamma t} - 1)/\gamma$. We remark that the exponent γ is essentially the same as the exponent in Theorem II.6.1.

It remains to prove the two lemmas.

The local error bound of Lemma 2.3 is proved by transfering the arguments of the proof of Lemma III.3.3 to the present non-linear setting via the calculus of Lie derivatives, which we describe next.

Calculus of Lie Derivatives. (Cf., e.g., Hairer, Lubich & Wanner (2006), Sect. III.5, or Hundsdorfer & Verwer (2003), Sect. IV.1.4. This formalism only relies on the differentiability and the semi-group property of the flow, and so it is applicable in the infinite-dimensional setting as well as in the finite-dimensional case.)

For a vector field F on $\mathcal{M}$, which to $u \in \mathcal{M}$ associates $F(u) \in \mathcal{T}_u\mathcal{M}$, such as $\widehat{T}$ or $\widehat{V}$ or $\widehat{H} = \widehat{T} + \widehat{V}$, we denote by ϕ_F^t the flow at time t of the differential equation $\dot{u} = F(u)$ on $\mathcal{M}$, that is, $\phi_F^t(v)$ is the solution at time t with initial value $u(0) = v$. The Lie derivative $D_F G$ at $v \in \mathcal{M}$ of another vector field G on $\mathcal{M}$ is defined by

$$(D_F G)(v) = \frac{d}{dt}\Big|_{t=0} G(\phi_F^t(v)) = G'(v)F(v).$$

We use the notation

$$\Big(\exp(tD_F)G\Big)(v) = G(\phi_F^t(v)).$$

In particular, for the identity Id, the flow is reproduced as $\exp(tD_F)\mathrm{Id}(v) = \phi_F^t(v)$. We then have the following properties:

$$\frac{d}{dt}\exp(tD_F)G(v) = \Big(\exp(tD_F)D_F G\Big)(v) = \Big(D_F \exp(tD_F)G\Big)(v).$$

The first equality follows directly from the definition. The second equality uses that $F(\phi_F^t(v)) = (\phi_F^t)'(v)F(v)$, which is obtained by observing that the difference $\delta(t) = F(\phi_F^t(v)) - (\phi_F^t)'(v)F(v)$ satisfies the linear differential equation $\dot{\delta}(t) = F'(\phi_F^t(v))\delta(t)$ with $\delta(0) = 0$.

The commutator $[D_F, D_G] = D_F D_G - D_G D_F$ of the Lie derivatives of two vector fields F and G is the Lie derivative of the Lie commutator $[G, F] = G'F - F'G$ of the vector fields in reversed order,

$$[D_F, D_G] = D_{[G,F]}.$$

This is seen by a direct calculation in which second derivatives cancel.

Proof of Lemma 2.3. For notational simplicity we write D_H, D_T, D_V instead of $D_{\widehat{H}}$, $D_{\widehat{T}}$, $D_{\widehat{V}}$, respectively. We start from the Gröbner–Alekseev or nonlinear variation-of-constants formula (see Hairer, Nørsett & Wanner, 1993, Section I.14), which in the present notation reads

$$\begin{aligned} u(\Delta t) = \exp(\Delta t\, D_H)\mathrm{Id}(u^0) &= \exp(\Delta t\, D_T)\mathrm{Id}(u^0) + \\ &\quad \int_0^{\Delta t} \exp((\Delta t - s)D_H)\, D_V\, \exp(sD_T)\mathrm{Id}(u^0)\, ds. \end{aligned}$$

Using this formula once more for the expression under the integral, we obtain

$$\begin{aligned} u(\Delta t) &= \exp(\Delta t\, D_T)\mathrm{Id}(u^0) + \\ &\quad \int_0^{\Delta t} \exp((\Delta t - s)D_T)\, D_V\, \exp(sD_T)\mathrm{Id}(u^0)\, ds + r_1 \end{aligned}$$

with the remainder

$$r_1 = \int_0^{\Delta t}\int_0^{\Delta t - s} \exp((\Delta t - s - \sigma)D_H) D_V \exp(\sigma D_T) D_V \exp(sD_T)\mathrm{Id}(u^0)\, d\sigma\, ds\;,$$

which is bounded by $C\Delta t^2$ with a constant C depending only on the bound β of the potential V. On the other hand, the numerical solution reads in this notation

$$u^1 = \exp(\frac{1}{2}\Delta t D_V)\exp(\Delta t D_T)\exp(\frac{1}{2}\Delta t D_V)\mathrm{Id}(u^0)\,.$$

Taylor expansion $\exp(\frac{\Delta t}{2}D_V) = I + \frac{\Delta t}{2}D_V + \left(\frac{\Delta t}{2}\right)^2 \int_0^1 (1-\theta)\exp(\theta\frac{\Delta t}{2}D_V)D_V^2\,d\theta$ gives

$$u^1 = \exp(\Delta t D_T)\mathrm{Id}(u^0) + \frac{\Delta t}{2}\big(\exp(\Delta t D_T)D_V + D_V\exp(\Delta t D_T)\big)\mathrm{Id}(u^0) + r_2$$

with the remainder r_2 bounded by $C\Delta t^2$, again with a constant C depending only on β. The error now becomes

$$\begin{aligned} u^1 - u(\Delta t) &= \frac{\Delta t}{2}\big(\exp(\Delta t D_T)D_V + D_V\exp(\Delta t D_T)\big)\mathrm{Id}(u^0) \\ &\quad - \int_0^{\Delta t} \exp((\Delta t - s)D_T)\,D_V\,\exp(sD_T)\mathrm{Id}(u^0)\,ds + (r_2 - r_1), \end{aligned}$$

and hence the principal error term is just the quadrature error of the trapezoidal rule applied to the integral over $[0,\Delta t]$ of the function

$$f(s) = \exp((\Delta t - s)D_T)\,D_V\,\exp(sD_T)\mathrm{Id}(u^0).$$

We express the quadrature error in first- and second-order Peano form,

$$\begin{aligned} &\frac{1}{2}\Delta t\,(f(0) + f(\Delta t)) - \int_0^{\Delta t} f(s)\,ds \\ &= -\Delta t^2 \int_0^1 (\frac{1}{2} - \theta)\,f'(\theta\Delta t)\,d\theta = \frac{1}{2}\Delta t^3 \int_0^1 \theta(1-\theta)f''(\theta\Delta t)\,d\theta\,. \end{aligned}$$

Since

$$\begin{aligned} f'(s) &= \exp((\Delta t - s)D_T)\,[D_T, D_V]\,\exp(sD_T)\mathrm{Id}(u^0) \\ &= -\exp((\Delta t - s)D_T)\,D_{[\widehat{T},\widehat{V}]}\,\exp(sD_T)\mathrm{Id}(u^0) \\ &= -e^{-isT}\,[\widehat{T},\widehat{V}]\,e^{-i(\Delta t - s)T}u^0\,, \end{aligned}$$

the commutator bound (2.9) shows that the quadrature error is bounded by $\frac{1}{4}c_1\Delta t^2\|u^0\|_1$. This proves the first error bound of Lemma 2.3. To obtain the second bound we use similarly

$$\begin{aligned} f''(s) &= \exp((\Delta t - s)D_T)\,[D_T,[D_T, D_V]]\,\exp(sD_T)\mathrm{Id}(u^0) \\ &= e^{-isT}\,[\widehat{T},[\widehat{T},\widehat{V}]]\,e^{-i(\Delta t - s)T}u^0\,, \end{aligned}$$

and hence (2.10) shows that the quadrature error is bounded by $\frac{1}{12} c_2 \, \Delta t^3 \, \|u^0\|_2$. A closer look at the remainder term $r = r_2 - r_1$ yields, as in the proof of Lemma III.3.3, that r is itself a quadrature error of a first-order two-dimensional quadrature formula for the integral over the triangle $0 \le s \le \Delta t$, $0 \le \sigma \le \Delta t - s$ of the function

$$g(s,\sigma) = \exp((\Delta t - s - \sigma) D_T)\, D_V \, \exp(\sigma D_T)\, D_V \, \exp(s D_T) \mathrm{Id}(u^0)\,,$$

plus an $\mathcal{O}(\Delta t^3)$ remainder term. The quadrature error in r is bounded by $C \Delta t^3$ times the norms of the partial derivatives with respect to s and σ of g, which turn out to be bounded as needed. □

Proof of Lemma 2.4. The stability estimate is obtained from (2.8) by observing that $e^{-i\Delta t T}$ is unitary and by showing that

$$\|\phi_V^t(u^0) - \phi_V^t(v^0)\| \le e^{\gamma t}\, \|u^0 - v^0\|$$

with γ as stated in the lemma. This bound is shown as follows: We write $u(t) = \phi_V^t(u^0)$ and $v(t) = \phi_V^t(v^0)$. Noting $u = P(u)u$ (recall the assumption $u \in \mathcal{T}_u\mathcal{M}$ for $u \in \mathcal{M}$) and $v = P(v)v$, we have

$$\dot u - \dot v = -iP(u)VP(u)(u-v) - i\big(P(u)VP(u) - P(v)VP(v)\big)v\,.$$

Forming the inner product with $u - v$ and taking the real part, we obtain

$$\begin{aligned}
\|u-v\| \cdot \frac{d}{dt}\|u-v\| &= \mathrm{Re}\,\langle u - v \,|\, \dot u - \dot v\rangle \\
&= \mathrm{Re}\,\big\langle u - v \,|\, -i\big(P(u)VP(u) - P(v)VP(v)\big)v\big\rangle \\
&= \ \mathrm{Im}\,\big\langle u - v \,|\, P(u)V\big(P(u) - P(v)\big)v\big\rangle \\
&\quad + \mathrm{Im}\,\big\langle u - v \,|\, \big(P(u) - P(v)\big)P(v)Vv\big\rangle \\
&\quad + \mathrm{Im}\,\big\langle u - v \,|\, \big(P(u) - P(v)\big)P^\perp(v)Vv\big\rangle \\
&\equiv I + II + III\,.
\end{aligned}$$

Since

$$\big(P(u) - P(v)\big)v = -\big(P^\perp(u) - P^\perp(v)\big)v = -P^\perp(u)v = P^\perp(u)(u - v)\,,$$

the bound (2.7) gives us

$$|I| \le \beta\kappa\, \|u - v\|^3\,.$$

For II we note

$$\begin{aligned}
&\big\langle u - v \,|\, \big(P(u) - P(v)\big)P(v)Vv\big\rangle = -\big\langle u - v \,|\, \big(P^\perp(u) - P^\perp(v)\big)P(v)Vv\big\rangle \\
&\quad = -\big\langle u - v \,|\, P^\perp(u)P(v)Vv\big\rangle = -\big\langle P^\perp(u)(u - v) \,|\, P^\perp(u)P(v)Vv\big\rangle \\
&\quad = \big\langle P^\perp(u)(u - v) \,|\, \big(P(u) - P(v)\big)P(v)Vv\big\rangle\,,
\end{aligned}$$

and hence (2.6)–(2.7) yield

$$|II| \leq \beta\kappa^2 \, \|u-v\|^3 .$$

Finally, since

$$\|P^{\perp}(v)Vv\| = \mathrm{dist}(Vv, \mathcal{T}_v\mathcal{M}) = \delta + \mathcal{O}(\Delta t) ,$$

we have from (2.6)

$$|III| \leq \kappa\big(\delta + \mathcal{O}(\Delta t)\big) \, \|u-v\|^2 .$$

Hence, as long as $\|u-v\| = \mathcal{O}(\Delta t)$, we obtain

$$\frac{d}{dt}\|u-v\| \leq \big(\kappa\delta + \mathcal{O}(\Delta t)\big) \, \|u-v\| ,$$

which yields the stated result. □

IV.3 Variational Splitting for MCTDH

We consider the Schrödinger equation for the nuclei obtained from the Born–Oppenheiner approximation,

$$i\hbar\frac{\partial\psi}{\partial t} = H\psi , \qquad H = T + V \tag{3.1}$$

with the kinetic energy operator $T = -\sum_{n=1}^{N} \frac{\hbar^2}{2M_N} \Delta_{x_n}$ and a potential $V(x_1, \dots, x_N)$, which we assume bounded and smooth for all theoretical statements made in this section.

We recall the MCTDH method of Sect. II.3.3, which gives the variational approximation on the manifold $\mathcal{M}$ of (II.3.40). The approximation is by linear combinations of tensor products of single-particle functions satisfying the orthogonality relations (II.3.31),

$$u = \sum_J a_J \, \varphi_{j_1}^{(1)} \otimes \cdots \otimes \varphi_{j_N}^{(N)} , \tag{3.2}$$

where the sum is over multi-indices $J = (j_1, \dots, j_N)$ for $j_n = 1, \dots, r_n$ and $n = 1, \dots, N$.

The variational splitting method for the MCTDH model turns out to yield an explicit time integration method that is unconditonally stable with respect to the space discretization parameter Δx. As an interesting alternative to the variational splitting method for integrating the MCTDH equations, we refer to the constant-mean-field multiple time-stepping method by Beck & Meyer (1997) where, however, the micro-steps are required of size $\mathcal{O}(\Delta x^2)$ for stability.

Step with the Kinetic Energy. Since $Tu \in \mathcal{T}_u\mathcal{M}$ for $u \in \mathcal{M} \cap D(T)$, the step (2.3) with T actually solves the free Schrödinger equation

$$i\hbar\,\frac{\partial u}{\partial t} = Tu .$$

For an initial function in the multi-configuration form (3.2), this completely decouples into single-particle free Schrödinger equations: $da_J/dt = 0$ for all J and

$$i\hbar \frac{\partial \varphi_{j_n}^{(n)}}{\partial t} = -\frac{\hbar^2}{2M_n} \Delta_{x_n} \varphi_{j_n}^{(n)} \tag{3.3}$$

for $j_n = 1, \dots, r_n$ and $n = 1, \dots, N$. These equations are solved efficiently, e.g., by FFT diagonalization of the Laplacian in the Fourier method of Section III.1.3. We note that the solutions of (3.3) preserve the orthogonality (3.31) of the single-particle functions.

Step with the Potential. Solving (2.2) means solving the MCTDH equations (II.3.41)–(II.3.42) with only the potential V instead of the full Hamiltonian H:

$$i\hbar \frac{da_J}{dt} = \sum_K \langle \Phi_J \,|\, V \,|\, \Phi_K \rangle \, a_K \,, \qquad \forall J = (j_1, \dots, j_N)\,, \tag{3.4}$$

$$i\hbar \frac{\partial \varphi_{j_n}^{(n)}}{\partial t} = (I - P^{(n)}) \sum_{k_n=1}^{r_n} \sum_{l_n=1}^{r_n} (\rho^{(n)})^{-1}_{j_n,k_n} \langle \psi_{k_n}^{(n)} \,|\, V \,|\, \psi_{l_n}^{(n)} \rangle^{(\neg n)} \, \varphi_{l_n}^{(n)}, \tag{3.5}$$

$$j_n = 1, \dots, r_n, \quad n = 1, \dots, N\,,$$

where we refer back to Theorem II.3.4 for the notation. The favourable situation compared with the full MCTDH equations is that no differential operator appears on the right-hand side of the equations. Therefore, with an explicit integrator the step size Δt can be chosen independent of the spatial grid size Δx. For the further discussion we collect all coefficients in $a = (a_J)$ and all single-particle functions in $\varphi = \big(\varphi_{j_n}^{(n)}\big)$ and abbreviate the differential equations (3.4)–(3.5) as

$$\begin{aligned} i\hbar \, \dot a &= A_V(\varphi)\, a \\ i\hbar \, \dot\varphi &= (I - P(\varphi))\, B_V(a, \varphi)\, \varphi\,. \end{aligned} \tag{3.6}$$

Here $A_V(\varphi)$ is the Galerkin matrix with entries $\langle \Phi_J \,|\, V \,|\, \Phi_K \rangle$, which are high-dimensional integrals that need to be computed approximately; see Beck, Jäckle, Worth & Meyer (2000) for various techniques employed in their MCTDH code. A highly successful approach is to approximate the potential by linear combinations of tensor products as in (II.3.13), for which the integrals reduce to products of low-dimensional integrals. In the second equation, $P = \mathrm{blockdiag}(P^{(k)})$ is the orthogonal projection and B_V contains the density matrices $\rho^{(n)}$ and the mean-field operators, which are formed by the inner products $\langle \psi_{k_n}^{(n)} \,|\, V \,|\, \psi_{l_n}^{(n)} \rangle^{(\neg n)}$ with the single-hole functions. The computation of the matrix elements is the computationally most expensive part of the method.

Time Integration Scheme. For the numerical solution of the differential equations (3.6) we consider a multiple time-stepping approach for the reduced MCTDH equations (3.6), a variant of the *constant mean-field* approximation of Beck & Meyer (1997) proposed there for the full MCTDH equations. This approach is motivated by the observation that the single-particle functions φ often change faster than the Galerkin matrix and the mean-field operators, whose computation is costly. We consider the following algorithm for integrating the reduced MCTDH equations (3.6).

Algorithm 3.1 (Constant Mean-Field Method for the Potential Part in MCTDH). *A time step of length Δt for the differential equations (3.6) proceeds as follows.*

1. *Compute $B_V^0 = B_V(a^0, \varphi^0)$ (that is, the density and mean-field matrices at (a^0, φ^0)) and solve*

$$i\hbar\dot{\varphi} = (I - P(\varphi))\, B_V^0\, \varphi \tag{3.7}$$

with initial value $\varphi(0) = \varphi^0$ over half a time step $[0, \frac{1}{2}\Delta t]$ and set $\varphi^{1/2} = \varphi(\frac{1}{2}\Delta t)$.

2. *Compute the Galerkin matrix $A_V^{1/2} = A_V(\varphi^{1/2})$ and*

$$a^1 = e^{-i\Delta t A_V^{1/2}/\hbar}\, a^0 .$$

3. *Determine the predictor $\widetilde{\varphi}^1$ as the solution of (3.7) at Δt and compute $B_V^1 = B_V(a^1, \widetilde{\varphi}^1)$. Solve*

$$i\hbar\dot{\varphi} = (I - P(\varphi))\, B_V^1\, \varphi \tag{3.8}$$

with initial value $\varphi(\frac{1}{2}\Delta t) = \varphi^{1/2}$ over $[\frac{1}{2}\Delta t, \Delta t]$ and set $\varphi^1 = \varphi(\Delta t)$.

The differential equations for φ in the first and third step are solved approximately by ν smaller time steps of length $\Delta\tau = \Delta t/\nu$. An orthogonality-preserving integrator is favourably used here; see Hairer, Lubich & Wanner (2006), Sect. IV.9.1, for various possibilities. The action of the matrix exponential in the second step is computed efficiently with the Lanczos method of Sect. III.2.2. There is no step size restriction by $\mathcal{O}(\Delta x^2)$ since the Laplacian of the kinetic energy part has been taken out by the variational splitting.

IV.4 Variational Splitting for Gaussian Wave Packets

We consider the Schrödinger equation (II.4.1) in semi-classical scaling, viz.,

$$i\varepsilon\frac{\partial\psi}{\partial t} = H\psi, \qquad H = H^\varepsilon = -\frac{\varepsilon^2}{2}\Delta + V, \tag{4.1}$$

with a small positive parameter ε. We recall from Sect. II.4 that the variational approximation of the wave function by a complex Gaussian (II.4.2), viz.,

$$\psi(x,t) \approx u(x,t) = \exp\Big(\frac{i}{\varepsilon}\Big(\frac{1}{2}(x-q(t))^T C(t)(x-q(t)) + p(t)\cdot(x-q(t)) + \zeta(t)\Big)\Big), \tag{4.2}$$

has equations of motion for the parameters that read, with averages $\langle W\rangle(t) = \langle u(t)\,|\,W\,|\,u(t)\rangle$,

$$\begin{aligned} \dot{q} &= p \\ \dot{p} &= -\langle\nabla V\rangle \end{aligned} \tag{4.3}$$

for position and momentum and

$$\dot{C} = -C^2 - \langle \nabla^2 V \rangle \tag{4.4}$$

$$\dot{\zeta} = \frac{1}{2}|p|^2 - \langle V \rangle + \frac{i\varepsilon}{2}\operatorname{tr} C + \frac{\varepsilon}{4}\left\langle \operatorname{tr}\left((\operatorname{Im} C)^{-1}\nabla^2 V\right)\right\rangle . \tag{4.5}$$

for the width matrix and phase.

Variational Splitting in Coordinates. The variational splitting method of Algorithm 2.1 applied to Gaussian wave-packet dynamics yields a corresponding splitting in the differential equations for the parameters:

1. and 3. *Half-step with V:*

$$\begin{cases} \dot{q} = 0 \\ \dot{p} = -\langle \nabla V \rangle \end{cases} \qquad \begin{cases} \dot{C} = -\langle \nabla^2 V \rangle \\ \dot{\zeta} = -\langle V \rangle + \frac{\varepsilon}{4}\left\langle \operatorname{tr}\left((\operatorname{Im} C)^{-1}\nabla^2 V\right)\right\rangle \end{cases} \tag{4.6}$$

2. *Full step with T:*

$$\begin{cases} \dot{q} = p \\ \dot{p} = 0 \end{cases} \qquad \begin{cases} \dot{C} = -C^2 \\ \dot{\zeta} = \frac{1}{2}|p|^2 + \frac{i\varepsilon}{2}\operatorname{tr} C \end{cases} \tag{4.7}$$

Remarkably, these differential equations can all be solved explicitly. We note that q, $\operatorname{Im} C$, and $\operatorname{Im}\zeta$ remain constant in time in (4.6), and hence also the averages of V, ∇V, and $\nabla^2 V$ over the Gaussian. Denoting by u^0 the Gaussian with parameters q^0, p^0, C^0, ζ^0 and $\langle \nabla W \rangle^0 = \langle u^0 \,|\, W \,|\, u^0 \rangle$ for $W = V, \nabla V, \nabla^2 V$, the solution of (4.6) is thus given by

$$\begin{cases} q(t) = q^0 \\ p(t) = p^0 - t\,\langle \nabla V \rangle^0 \end{cases} \qquad \begin{cases} C(t) = C^0 - \frac{t}{2}\langle \nabla^2 V \rangle^0 \\ \zeta(t) = \zeta^0 - t\,\langle V \rangle^0 + \frac{t\varepsilon}{8}\left\langle \operatorname{tr}\left((\operatorname{Im} C^0)^{-1}\nabla^2 V\right)\right\rangle^0 . \end{cases}$$

The solution to the differential equations (4.7) is obtained as

$$\begin{cases} q(t) = q^0 + tp^0 \\ p(t) = p^0 \end{cases} \qquad \begin{cases} C(t) = C^0\,(I + tC^0)^{-1} \\ \zeta(t) = \zeta^0 + \frac{t}{2}\,|p^0|^2 + \frac{i}{2}\,\varepsilon \operatorname{tr}\bigl(\log(I + tC^0)\bigr). \end{cases}$$

This yields the following numerical method from Faou & Lubich (2006).

Algorithm 4.1 (Gaussian Wave-Packet Integrator). *Starting from the Gaussian u^n with parameters q^n, p^n, C^n, ζ^n, a time step for (4.3)–(4.5) from time t^n to $t^{n+1} = t^n + \Delta t$ proceeds as follows:*

1. With the averages $\langle W \rangle^n = \langle u^n \,|\, W \,|\, u^n \rangle$ for $W = V, \nabla V, \nabla^2 V$, compute

$$\begin{aligned} p^{n+1/2} &= p^n - \frac{\Delta t}{2}\,\langle \nabla V \rangle^n \\ C_+^n &= C^n - \frac{\Delta t}{2}\,\langle \nabla^2 V \rangle^n \\ \zeta_+^n &= \zeta^n - \frac{\Delta t}{2}\,\langle V \rangle^n + \frac{\Delta t\,\varepsilon}{8}\left\langle \operatorname{tr}\left((\operatorname{Im} C^n)^{-1}\nabla^2 V\right)\right\rangle^n . \end{aligned} \tag{4.8}$$

2. *Compute*

$$\begin{aligned} q^{n+1} &= q^n - \Delta t\, p^{n+1/2} \\ C_-^{n+1} &= C_+^n (I + \Delta t\, C_+^n)^{-1} \\ \zeta_-^{n+1} &= \zeta_+^n + \frac{\Delta t}{2} \left|p^{n+1/2}\right|^2 + \frac{i}{2}\, \varepsilon \operatorname{tr}\bigl(\log(I + \Delta t\, C_+^n)\bigr)\,. \end{aligned} \tag{4.9}$$

3. *With the averages over the Gaussian at time t^{n+1}, which are the same as those for the previously computed parameters q^{n+1}, C_-^{n+1}, ζ_-^{n+1}, compute*

$$\begin{aligned} p^{n+1} &= p^{n+1/2} - \frac{\Delta t}{2} \langle \nabla V \rangle^{n+1} \\ C^{n+1} &= C_-^{n+1} - \frac{\Delta t}{2} \langle \nabla^2 V \rangle^{n+1} \\ \zeta^{n+1} &= \zeta_-^{n+1} - \frac{\Delta t}{2} \langle V \rangle^{n+1} + \frac{\Delta t\, \varepsilon}{8} \Bigl\langle \operatorname{tr}\bigl((\operatorname{Im} C^{n+1})^{-1} \nabla^2 V\bigr) \Bigr\rangle^{n+1}. \end{aligned} \tag{4.10}$$

We collect some properties of this algorithm.

Theorem 4.2 (Properties of the Gaussian Wave-Packet Integrator). *Algorithm 4.1 is an explicit, second-order numerical method for Gaussian wave-packet dynamics. The method is symplectic and time-reversible and preserves the unit L^2 norm of the wave packets u^n. In the limit $\varepsilon \to 0$, the position and momentum approximations q^n, p^n of this method tend to those obtained by applying the Störmer–Verlet method (I.3.5) to the associated classical mechanical system (I.1.1).*

Proof. The statement for $\varepsilon \to 0$ follows directly from the equations for $p^{n+1/2}$, q^{n+1}, p^{n+1} and from noting $\langle \nabla V \rangle^n \to \nabla V(q^n)$. The other properties have already been verified for variational splitting methods (Algorithm 2.1) in general. □

The method does not preserve the total energy exactly, but it is shown by Faou & Lubich (2006) that the energy $\langle u^n \mid H \mid u^n \rangle$ along the numerical solution deviates from the initial energy only by $\mathcal{O}(\Delta t^2)$ over exponentially long times $t \le e^{c/\Delta t}$, uniformly in ε. The proof uses the symplecticity of the method, in the form of the preservation of the Poisson structure (see Sect. II.4.2) by the one-step map for the parameters.

In view of the small parameter ε, the discussion of the order of the method requires some care. Here it is useful to consider the integrator in the scaled variables $\widehat{y} = (p, q, \operatorname{Re} C, \operatorname{Im} C/\varepsilon, \operatorname{Re} \zeta, \operatorname{Im} \zeta/\varepsilon)$. Since the equations of motion in the scaled variables turn out to contain ε only as a regular perturbation parameter, after n steps of the splitting integrator we have the ε-uniform error bound

$$\widehat{y}^n - \widehat{y}(t^n) = \mathcal{O}(\Delta t^2),$$

where the constants symbolized by the $\mathcal{O}$-notation are independent of ε and of n and Δt with $n\Delta t \le \mathrm{Const}$. For the absolute values of the Gaussian wave packets this yields the error bound

$$\big\| |u^n|^2 - |u(t^n)|^2 \big\| = \mathcal{O}(\Delta t^2), \tag{4.11}$$

but the approximation including the phase is only

$$\|u^n - u(t^n)\| = \mathcal{O}(\Delta t^2/\varepsilon). \tag{4.12}$$

We refer to Faou & Lubich (2006) for more details, for the formulation of the algorithm for spherical wave packets (diagonal complex width matrix C), and for numerical experiments that illustrate the stated properties.

Chapter V. Semi-Classical Dynamics Using Hagedorn Wave Packets

The time-dependent Schrödinger equation in semi-classical scaling,

$$i\varepsilon\frac{\partial\psi}{\partial t} = -\frac{\varepsilon^2}{2m}\Delta\psi + V\psi \tag{0.1}$$

with a small parameter ε, has typical solutions that are wavepackets of width $\sim \sqrt{\varepsilon}$, highly oscillatory with wavelength $\sim \varepsilon$, and with the envelope moving at velocity ~ 1. Grid-based numerical methods therefore need very fine resolution for small ε and thus become computationally infeasible or at least very expensive; cf. Markowich, Pietra & Pohl (1999). The highly oscillatory solution behaviour also excludes the approximation of the wave function on sparse grids in higher dimensions, because the necessary smoothness requirements for this technique are not met for small ε; see Gradinaru (2008).

In this chapter we describe a numerical approach that is robust in the semi-classical limit $\varepsilon \to 0$. The wave function is approximated by moving and deforming complex Gaussians times polynomials, in a suitable parameter-dependent orthonormal basis developed by Hagedorn (1998).

V.1 Hagedorn's Parametrization of Gaussian Wave Packets

In Sect. II.4 we considered Gaussian wave packets written as

$$u(x,t) = \exp\left(\frac{i}{\varepsilon}\left(\frac{1}{2}(x-q(t))^T C(t)(x-q(t)) + p(t)\cdot(x-q(t)) + \zeta(t)\right)\right),$$

with a d-dimensional complex symmetric matrix $C(t)$ with positive definite imaginary part. It was noticed by Hagedorn (1980) and further developed in subsequent papers, notably in Hagedorn (1998), that much insight and important extensions can be obtained from factorizing $C(t)$ into two complex matrices with special properties.

A Matrix Factorization. A key to the further development is the following matrix lemma, see Hagedorn (1998), Sect. 3. Here, the superscript T denotes the transpose of a matrix and the superscript $*$ denotes the transpose and complex conjugate. I is the d-dimensional identity matrix. (The matrices Q and P in the lemma correspond to A and iB in Hagedorn's papers.)

Lemma 1.1. *Let Q and P be complex $d \times d$ matrices that satisfy the relations*

$$\begin{aligned} Q^T P - P^T Q &= 0 \\ Q^* P - P^* Q &= 2iI \,. \end{aligned} \tag{1.1}$$

Then, Q and P are invertible, and

$$C = PQ^{-1}$$

is complex symmetric with the positive definite imaginary part

$$\operatorname{Im} C = (QQ^*)^{-1}. \tag{1.2}$$

Conversely, every complex symmetric matrix C with positive definite imaginary part can be written as $C = PQ^{-1}$ with matrices Q and P satisfying (1.1).

Proof. Multiplying the second equation of (1.1) from the left and the right with a vector $v \in \mathbb{C}^d$ yields

$$(Qv)^*(Pv) - (Pv)^*(Qv) = 2i \, \|v\|^2 \,,$$

which shows that $v = 0$ is the only vector in the null-space of Q or P. Hence, these matrices are invertible. Multiplying the first equation of (1.1) from the left with $(Q^{-1})^T$ and from the right with Q^{-1} gives

$$PQ^{-1} - (Q^{-1})^T P^T = 0$$

and thus shows that $C = PQ^{-1}$ is complex symmetric. Further, we have

$$(\operatorname{Im} C)(QQ^*) = \frac{1}{2i}(PQ^{-1} - (Q^{-1})^* P^*)QQ^* = \frac{1}{2i}(PQ^* - (Q^*)^{-1}(P^*Q)Q^*) \,,$$

which simplifies to the identity on using the second equation of (1.1) for substituting $P^*Q = Q^*P - 2iI$.

Conversely, for a complex symmetric matrix C with positive definite imaginary part we set $Q = (\operatorname{Im} C)^{-1/2}$ and $P = CQ$. It is readily verified that these matrices satisfy the relations (1.1). □

The factorization is not unique, since multiplying Q and P from the right with a unitary matrix preserves the relations (1.1).

Relationship with Symplectic Matrices. A real matrix $Y \in \mathbb{R}^{2d \times 2d}$ is *symplectic* if it satisfies the quadratic relation

$$Y^T J Y = J \quad \text{with} \quad J = \begin{pmatrix} 0 & -I \\ I & 0 \end{pmatrix}. \tag{1.3}$$

If we set, for complex matrices $Q, P \in \mathbb{C}^{d \times d}$,

$$Y = \begin{pmatrix} \operatorname{Re} Q & \operatorname{Im} Q \\ \operatorname{Re} P & \operatorname{Im} P \end{pmatrix},$$

then the relations (1.1) are equivalent to the symplecticity condition (1.3). We shall therefore refer to (1.1) as the *symplecticity relations* for Q and P.

Complex Gaussians in Hagedorn's Parametrization. Consider a normalized d-dimensional Gaussian

$$\varphi_0^\varepsilon[q,p,Q,P](x) = (\pi\varepsilon)^{-d/4}(\det Q)^{-1/2}\exp\Big(\frac{i}{2\varepsilon}(x-q)^T PQ^{-1}(x-q)+\frac{i}{\varepsilon}p^T(x-q)\Big) \tag{1.4}$$

with $q,p \in \mathbb{R}^d$ and matrices $Q,P \in \mathbb{C}^{d\times d}$ satisfying (1.1). This function is of unit L^2 norm since a diagonalization of $\operatorname{Im} PQ^{-1}$ shows that it should have the factor $|\det(\operatorname{Im} PQ^{-1})^{-1}|^{-1/4}$, and by (1.2), $|\det(\operatorname{Im} PQ^{-1})^{-1}|^{-1/4} = |\det(QQ^*)|^{-1/4} = |\det Q|^{-1/2}$. It turns out favourable to take $(\det Q)^{-1/2}$ without the absolute value, where the branch of the square root is chosen suitably, in particular such that $(\det Q(t))^{-1/2}$ is a continuous function of t for a continuous family of invertible matrices $Q(t)$. We write φ_0 instead of $\varphi_0^\varepsilon[q,p,Q,P]$ when the parameters are clear from the context.

Gaussian Wave Packets and Quadratic Hamiltonians. We know from Sect. II.4 that Gaussian wave packets with appropriately evolving parameters are exact solutions to time-dependent Schrödinger equations (0.1) in the case of a *quadratic* potential V. This fact underlies the approximation result for more general smooth potentials V, see Theorem II.4.4. It turns out that the equations of motion assume a particularly appealing form with Hagedorn's parametrization.

We consider the classical equations of motion associated with (0.1),

$$\begin{aligned} \dot q &= \frac{p}{m} \\ \dot p &= -\nabla V(q) \end{aligned} \tag{1.5}$$

and their linearization along $(q(t),p(t))$,

$$\begin{aligned} \dot Q &= \frac{P}{m} \\ \dot P &= -\nabla^2 V(q) Q \end{aligned} \tag{1.6}$$

with the Hessian matrix $\nabla^2 V(q)$. Further, we consider the classical action integral

$$S(t) = \int_0^t \Big(\frac{|p(s)|^2}{2m} - V(q(s))\Big)\,ds. \tag{1.7}$$

Then, there is the following basic result.

Theorem 1.2 (Gaussian Wave Packets in a Quadratic Potential, Hagedorn 1980). *Let V be a quadratic potential, and let $(q(t), p(t), Q(t), P(t))$ for $0 \le t \le \bar{t}$ be a solution of the classical equations (1.5)–(1.6) and $S(t)$ the corresponding action (1.7). Assume that $Q(0)$ and $P(0)$ satisfy the relations (1.1). Then, $Q(t)$ and $P(t)$ satisfy (1.1) for all times t, and*

$$\psi(x,t) = e^{iS(t)/\varepsilon}\, \varphi_0^\varepsilon[q(t), p(t), Q(t), P(t)](x) \tag{1.8}$$

is a solution of the time-dependent Schrödinger equation (0.1).

Proof. The fact that relation (1.1) is preserved under (1.6), is a consequence of the lemma below (or of the symplecticity of the flow of the Hamiltonian system (1.5)). A direct, lengthy calculation shows that (1.8) is a solution of the Schrödinger equation (0.1). Here it is useful to note that $\frac{d}{dt}(\det Q) = \mathrm{tr}(\dot{Q}Q^{-1}) = \frac{1}{m}\,\mathrm{tr}(PQ^{-1})$. □

Remark 1.3. The theorem remains valid also in the case of a time-dependent quadratic potential $V(x,t)$, in particular, for the local quadratic approximation to a smooth potential along a classical trajectory.

Lemma 1.4. *Suppose that $Q(t), P(t) \in \mathbb{C}^{d\times d}$ satisfy the differential equations*

$$\begin{aligned} \dot{Q}(t) &= F(t)P(t) \\ \dot{P}(t) &= G(t)Q(t) \end{aligned}$$

with real symmetric matrices $F(t)$, $G(t)$. If the relations (1.1) hold at $t = 0$, then they hold for all t.

Proof. We have

$$\frac{d}{dt}(Q^*P - P^*Q) = \dot{Q}^*P + Q^*\dot{P} - \dot{P}^*Q - P^*\dot{Q} = P^*FP + Q^*GQ - Q^*GQ - P^*FP = 0$$

and in the same way $\frac{d}{dt}(Q^TP - P^TQ) = 0$. □

Fourier Transform. Hagedorn's parametrization of Gaussians has further beautiful properties. We mention that the scaled Fourier transform

$$\mathcal{F}_\varepsilon\varphi(\xi) = (2\pi\varepsilon)^{-d/2} \int_{\mathbb{R}^d} \varphi(\xi)\, e^{-i\xi\cdot x/\varepsilon}\, d\xi$$

of the Gaussian φ_0^ε is given by the formula

$$\mathcal{F}_\varepsilon\varphi_0^\varepsilon[q,p,Q,P](\xi) = e^{-ip\cdot q/\varepsilon}\varphi_0^\varepsilon[p,-q,P,-Q](\xi)\,; \tag{1.9}$$

see Hagedorn (1998), formula (3.19).

V.2 Hagedorn's Semi-Classical Wave Packets

Following Hagedorn (1998), we construct parameter-dependent orthonormal $L^2(\mathbb{R}^d)$ bases of multi-variate polynomials times Gaussians, which have very favourable propagation properties in the time-dependent Schrödinger equation (0.1). The Hagedorn functions reduce to shifted and scaled Hermite functions in the one-dimensional case, but cannot in general be reduced to tensor products of Hermite functions in higher dimensions.

Ladder Operators. In Sect. III.1 we constructed the Hermite functions via the ladder operators for the harmonic oscillator. An analogous construction via appropriate parameter-dependent ladder operators yields the Hagedorn functions.

As in the previous section, we let $\varepsilon > 0$ be the small semi-classical parameter in (0.1). We let $q, p \in \mathbb{R}^d$ position and momentum parameters, and $Q, P \in \mathbb{C}^{d\times d}$ complex matrices satisfying the symplecticity relations (1.1). In this section, we denote the position and momentum *operators* by $\widehat{q} = (\widehat{q}_j)_{j=1}^d$ and $\widehat{p} = (\widehat{p}_j)_{j=1}^d$, respectively: for $\psi \in \mathcal{S}(\mathbb{R}^d)$,

$$(\widehat{q}\psi)(x) = x\psi(x), \qquad (\widehat{p}\psi)(x) = -i\varepsilon\nabla\psi(x) \qquad (x \in \mathbb{R}^d).$$

Compared with our previous notation, the hats are added to avoid confusion with the Gaussian parameters q and p. We recall the commutator relation (I.4.8), viz.,

$$\frac{1}{i\varepsilon}\,[\widehat{q}_j, \widehat{p}_k] = \delta_{jk}\,. \tag{2.1}$$

Hagedorn (1998) introduces the parameter-dependent *ladder operators* $A = (A_j)_{j=1}^d$ and $A^\dagger = (A_j^\dagger)_{j=1}^d$ as

$$\begin{aligned} A\ &= A[q,p,Q,P] = -\frac{i}{\sqrt{2\varepsilon}}\Big(P^T(\widehat{q}-q) - Q^T(\widehat{p}-p)\Big) \\ A^\dagger &= A^\dagger[q,p,Q,P] = \frac{i}{\sqrt{2\varepsilon}}\Big(P^*(\widehat{q}-q) - Q^*(\widehat{p}-p)\Big)\,. \end{aligned} \tag{2.2}$$

We note that for $d = 1$, $\varepsilon = 1$, $q = 0$, $p = 0$, $Q = 1$, $P = i$ we have again the ladder operators (III.1.6) of the standard harmonic oscillator. The key properties (III.1.8) and (III.1.9) extend as follows.

Lemma 2.1. *If Q and P satisfy the symplecticity relations (1.1), then we have the commutator identities*

$$[A_j, A_k^\dagger] = \delta_{jk}\,. \tag{2.3}$$

Moreover, $A_j^\dagger$ is adjoint to A_j on the Schwartz space $\mathcal{S}$:

$$\langle A_j^\dagger\varphi\,|\,\psi\rangle = \langle\varphi\,|\,A_j\psi\rangle \qquad \forall\,\varphi,\psi \in \mathcal{S}\,. \tag{2.4}$$

Proof. (a) With $Q = (Q_{jk})$ and $P = (P_{jk})$, we have (we let $q = p = 0$ for simplicity)

$$[A_j, A_k^\dagger] = \frac{1}{2\varepsilon}\Big[\sum_{\ell=1}^d (P_{\ell j}\widehat{q}_\ell - Q_{\ell j}\widehat{p}_\ell)\,,\ \sum_{m=1}^d (\overline{P}_{mk}\widehat{q}_m - \overline{Q}_{mk}\widehat{p}_m)\Big]\,.$$

By the canonical commutator relations (2.1), this simplifies to

$$[A_j, A_k^\dagger] = \frac{i}{2}\sum_{\ell=1}^{d}(-P_{\ell j}\overline{Q}_{\ell k} + Q_{\ell j}\overline{P}_{\ell k}) = \frac{i}{2}\big(-Q^*P + P^*Q\big)_{k,j}\,.$$

By (1.1), this equals δ_{jk}.

(b) To verify (2.4), we write out

$$\begin{aligned}\langle A_j^\dagger\varphi\,|\,\psi\rangle &= \Big\langle \frac{i}{\sqrt{2\varepsilon}}\sum_{\ell=1}^{d}(\overline{P}_{\ell j}\widehat{q}_\ell - \overline{Q}_{\ell j}\widehat{p}_\ell)\varphi\,\Big|\,\psi\Big\rangle \\ &= \Big\langle\varphi\,\Big|\,-\frac{i}{\sqrt{2\varepsilon}}\sum_{\ell=1}^{d}(P_{\ell j}\widehat{q}_\ell - Q_{\ell j}\widehat{p}_\ell)\psi\Big\rangle = \langle\varphi\,|\,A_j\psi\rangle\,,\end{aligned}$$

where we just use that $\widehat{q}_\ell$ and $\widehat{p}_\ell$ are self-adjoint operators. □

Lemma 2.2. *If Q and P satisfy the symplecticity relations (1.1), then the complex Gaussian $\varphi_0 = \varphi_0^\varepsilon[q,p,Q,P]$ of (1.4) spans the null-space of A.*

Proof. If $\varphi \in \mathcal{S}$ is in the null-space of A, then it must satisfy the linear system of partial differential equations

$$-i\varepsilon\nabla\varphi(x) - p\varphi(x) = C(x-q)\varphi(x)$$

with the complex symmetric matrix $C = PQ^{-1}$. Multiples of φ_0 are the only non-trivial solutions of this equation. □

Hagedorn Functions. In the same way as for the harmonic oscillator eigenfunctions in Sect. III.1, we can now construct eigenfunctions of the operators $A_jA_j^\dagger$ to eigenvalues $1, 2, 3, \dots$. Let $k = (k_1, \dots, k_d)$ be a multi-index with non-negative integers k_j, and denote by $\langle j\rangle = (0\dots1\dots0)$ the jth d-dimensional unit vector. By the same reasoning as for (III.1.11) we define recursively functions $\varphi_k = \varphi_k^\varepsilon[q,p,Q,P]$ by

$$\varphi_{k+\langle j\rangle} = \frac{1}{\sqrt{k_j+1}}\,A_j^\dagger\varphi_k \tag{2.5}$$

and find that the φ_k are normalized eigenfunctions of the symmetric operators $A_jA_j^\dagger$:

$$A_jA_j^\dagger\varphi_k = (k_j+1)\varphi_k, \qquad \|\varphi_k\| = 1\,.$$

From this relation we obtain, in the same way as in (III.1.12),

$$\varphi_{k-\langle j\rangle} = \frac{1}{\sqrt{k_j}}\,A_j\varphi_k\,, \tag{2.6}$$

so that $A_j^\dagger$ and A_j can be viewed as raising and lowering operators, respectively, in the jth component of the multi-index. From (2.5) and (2.6), using the definitions (2.2) and the fact that QQ^* is a real matrix by (1.2), we obtain the recurrence relation

$$Q\Big(\sqrt{k_j+1}\,\varphi_{k+\langle j\rangle}(x)\Big)_{j=1}^{d} = \sqrt{\frac{2}{\varepsilon}}\,(x-q)\varphi_k(x) - \overline{Q}\Big(\sqrt{k_j}\,\varphi_{k-\langle j\rangle}(x)\Big)_{j=1}^{d}. \qquad (2.7)$$

This permits us to evaluate $\varphi_k(x)$ at a given point x. It also shows that φ_k is the product of a polynomial of degree $k_1 + \ldots + k_d$ with the Gaussian φ_0.

Theorem 2.3 (Hagedorn Functions, Hagedorn 1998). *The functions $\varphi_k = \varphi_k^\varepsilon[q,p,Q,P]$ defined by (1.4) and (2.5) form a complete L^2-orthonormal set of functions.*

The orthonormality of the functions φ_k follows from their property as eigenfunctions of the symmetric operator $AA^\dagger$. The completeness is obtained by an extension of the arguments in the proof of completeness of the Hermite functions and is not proved here.

We mention that formula (1.9) for the Fourier transform extends to all the functions φ_k:

$$\mathcal{F}_\varepsilon \varphi_k^\varepsilon[q,p,Q,P](\xi) = (-i)^{|k|}\, e^{-ip\cdot q/\varepsilon}\, \varphi_k^\varepsilon[p,-q,P,-Q](\xi) \qquad (2.8)$$

with $|k| = k_1 + \ldots + k_d$; see Hagedorn (1998).

Evolution of Ladder Operators Under Quadratic Hamiltonians. Along a solution $(q(t),p(t),Q(t),P(t))$ of the classical equations (1.5)–(1.6) we consider the time-dependent operators

$$A_j(t) = A_j[q(t),p(t),Q(t),P(t)], \qquad A_j^\dagger(t) = A_j^\dagger[q(t),p(t),Q(t),P(t)]\,.$$

Let $H = -\frac{\varepsilon^2}{2m}\Delta + V$ denote the Hamiltonian of (0.1). The (non-self-adjoint) ladder operators evolve according to equations of the form of the Heisenberg equation (I.4.4).

Lemma 2.4. *In the case of a quadratic potential V we have*

$$\dot{A}_j = \frac{1}{i\varepsilon}[A_j,H]\,, \qquad \dot{A}_j^\dagger = -\frac{1}{i\varepsilon}[A_j^\dagger,H]\,.$$

Proof. With (1.5)–(1.6) we obtain for $A(t) = (A_j(t))$

$$\dot{A} = \frac{i}{\sqrt{2\varepsilon}}\big(Q^T\nabla V(\widehat{q}) + \frac{1}{m}P^T\widehat{p}\big).$$

The same expression is obtained for $\frac{1}{i\varepsilon}[A,H]$ on using the commutator relations (2.1) and the ensuing $\frac{1}{i\varepsilon}[\widehat{q}_j,\widehat{p}_k^2] = \delta_{jk}\cdot 2\widehat{p}_k$ and $\frac{1}{i\varepsilon}[\widehat{q}_j^2,\widehat{p}_k] = \delta_{jk}\cdot 2\widehat{q}_j$. The result for $A^\dagger$ is obtained by taking complex conjugates. □

Hagedorn Wave Packets and Quadratic Hamiltonians. We now have all ingredients for the following remarkable result.

Theorem 2.5 (Hagedorn Functions in a Quadratic Potential, Hagedorn 1998). *Let V be a quadratic potential, and let $(q(t), p(t), Q(t), P(t))$ be a solution of the classical equations (1.5)–(1.6) and $S(t)$ the corresponding action (1.7). Assume that $Q(0)$ and $P(0)$ satisfy the symplecticity relations (1.1). Then, for every multi-index k,*

$$e^{iS(t)/\varepsilon} \varphi_k^\varepsilon[q(t), p(t), Q(t), P(t)](x)$$

is a solution of the time-dependent Schrödinger equation (0.1).

Proof. We know from Theorem 1.2 that the statement is correct for $k = 0$. In view of the construction of the functions φ_k by (2.5), the result follows by induction if we can show that with a solution $\psi(\cdot, t)$, also $A_j^\dagger(t)\psi(\cdot, t)$ is a solution of (0.1). This holds indeed true, because

$$i\varepsilon \frac{\partial}{\partial t}(A_j^\dagger \psi) = i\varepsilon \dot{A}_j^\dagger \psi + A_j^\dagger H\psi = \Big(i\varepsilon \dot{A}_j^\dagger \psi + [A_j^\dagger, H]\psi\Big) + HA_j^\dagger \psi$$

and the expression in big brackets vanishes by Lemma 2.4. □

As with Theorem 1.2, the above theorem remains valid, with the same proof, for a time-dependent quadratic potential $V(x, t)$.

Approximation Schemes in the Case of Non-Quadratic Potentials. For a time-dependent Schrödinger equation (0.1) with a non-quadratic potential V, the wave function can be represented in the basis of Hagedorn functions as a series

$$\psi(x, t) = e^{iS(t)/\varepsilon} \sum_{k \in \mathbb{N}^d} a_k(t)\, \varphi_k(x, t) \tag{2.9}$$

(with $\mathbb{N}$ denoting the set of natural numbers including 0), where we have abbreviated $\varphi_k(x, t) = \varphi_k^\varepsilon[q(t), p(t), Q(t), P(t)](x)$ with $(q(t), p(t), Q(t), P(t))$ a solution to the classical equations (1.5)–(1.6) and where $S(t)$ is the classical action (1.7). We search for an approximation

$$\psi(x, t) \approx \psi_\mathcal{K}(x, t) = e^{iS(t)/\varepsilon} \sum_{k \in \mathcal{K}} c_k(t)\, \varphi_k(x, t) \tag{2.10}$$

with a finite multi-index set $\mathcal{K}$, which may be a cube $\{|k_j| \le K\}$ or a hyperbolically reduced set (III.1.18).

Hagedorn (1998) determines the coefficients c_k from differential equations that contain higher derivatives (from the third derivative onwards) of the potential V at the classical position $q(t)$. He obtains approximations to the wave function of asymptotic order $\mathcal{O}(\varepsilon^{N/2})$ for arbitrary N.

Alternatively, as in Sect. II.5.2, the coefficients can be determined from the time-dependent variational principle on the time-varying approximation space spanned by the functions $\varphi_k(t) = \varphi_k(\cdot, t)$ for $k \in \mathcal{K}$, by the Galerkin condition

$$\Big\langle \varphi_k(t) \,\Big|\, i\varepsilon \frac{\partial \psi_\mathcal{K}}{\partial t}(t) - H\psi_\mathcal{K}(t) \Big\rangle = 0 \qquad \forall\, k \in \mathcal{K}, \ \forall\, t\,.$$

If we write the potential as

$$V = U_{q(t)} + W_{q(t)}$$

with the quadratic Taylor polynomial U_q of V at the position q and with the non-quadratic remainder W_q, then we have

$$\begin{aligned} i\varepsilon\frac{\partial}{\partial t}(c_k e^{iS/\varepsilon}\varphi_k) - H(c_k e^{iS/\varepsilon}\varphi_k) &= i\varepsilon\dot{c}_k e^{iS/\varepsilon}\varphi_k \\ +c_k\Big(i\varepsilon\frac{\partial}{\partial t}(e^{iS/\varepsilon}\varphi_k) + \frac{\varepsilon^2}{2m}\Delta(e^{iS/\varepsilon}\varphi_k) &- U_q(e^{iS/\varepsilon}\varphi_k)\Big) - c_k e^{iS/\varepsilon} W_q \varphi_k \end{aligned}$$

where the term in big brackets vanishes by the version of Theorem 2.5 for time-dependent quadratic potentials. (Note that only the quadratic part U_q enters into the equations determining $q(t), p(t), Q(t), P(t), S(t)$). We then obtain the differential equations for the coefficients $c = (c_k)_{k\in\mathcal{K}}$ as

$$i\varepsilon\dot{c}(t) = F(t)c(t) \qquad \text{with} \quad F(t) = \big(\langle\varphi_k \,|\, W_{q(t)} \,|\, \varphi_\ell\rangle\big)_{k,\ell\in\mathcal{K}}.$$

In the following section we give a fully discrete, explicit, and time-reversible time-stepping algorithm to propagate the Gaussian parameters $q(t), p(t), Q(t), P(t)$, the phase $S(t)$, and the coefficients $c_k(t)$ of a Hagedorn wave packet (2.10).

V.3 A Numerical Integrator for Hagedorn Wave Packets

We describe an algorithm by Faou, Gradinaru & Lubich (2008) for the approximate solution of time-dependent Schrödinger equations (0.1) in the semi-classical regime using Hagedorn wave packets. The method is based on the splitting between the kinetic and potential operators $T = -\frac{\varepsilon^2}{2m}\Delta$ and V. We consider the free linear Schrödinger equation

$$i\varepsilon\frac{\partial\psi}{\partial t} = -\frac{\varepsilon^2}{2m}\Delta\psi \tag{3.1}$$

and the equation with only a potential

$$i\varepsilon\frac{\partial\psi}{\partial t} = V(x)\psi. \tag{3.2}$$

The potential will be further decomposed into its quadratic part at the current position q and the non-quadratic remainder.

Starting with a Hagedorn wave packet (2.10) as initial data for the Schrödinger equation, we will make use of the following:

- We can solve exactly the free Schrödinger equation (3.1). The wave function remains a Hagedorn wave packet (2.10) with unaltered coefficients c_k.
- For a quadratic potential, we can solve exactly equation (3.2). The wave function again remains a Hagedorn wave packet (2.10) with unaltered coefficients c_k.

- For the non-quadratic remainder at the current position, we compute the Galerkin approximation to equation (3.2) on the space spanned by the Hagedorn functions φ_k with fixed parameters q, p, Q, P, letting the coefficients c_k of (2.10) vary.

Kinetic Part and Quadratic Potential. We will use the following properties, which are direct consequences of Theorem 2.5. A time-dependent Hagedorn wavepacket (2.10) solves the free Schrödinger equation (3.1) if

$$\begin{aligned} q(t) &= q(0) + \frac{t}{m}\,p(0) \\ Q(t) &= Q(0) + \frac{t}{m}\,P(0) \\ S(t) &= S(0) + \frac{t}{2m}\,|p(0)|^2 \end{aligned} \tag{3.3}$$

and $p(t) = p(0)$, $P(t) = P(0)$, $c_k(t) = c_k(0)$.

For a quadratic potential $U(x)$, a time-dependent Hagedorn wavepacket (2.10) solves equation (3.2) with $V = U$ if

$$\begin{aligned} p(t) &= p(0) - t\,\nabla U(q(0)) \\ P(t) &= P(0) - t\,\nabla^2 U(q(0))Q(0) \\ S(t) &= S(0) - t\,U(q(0)) \end{aligned} \tag{3.4}$$

and $q(t) = q(0)$, $Q(t) = Q(0)$, $c_k(t) = c_k(0)$.

Galerkin Approximation for Non-Quadratic Potentials. Let $W(x)$ be a given (non-quadratic) potential and consider equation (3.2) with W in place of V. We fix the Gaussian parameters q, p, Q, P and take the Galerkin approximation on the linear space spanned by the Hagedorn functions $\varphi_k = \varphi_k^\varepsilon[q, p, Q, P]$ for multi-indices $k \in \mathcal{K}$. As we know from the beginning of Sect. III.1.1, this is equivalent to the linear system of ordinary differential equations for the coefficients $c(t) = (c_k(t))_{k\in\mathcal{K}}$ of (2.10):

$$i\varepsilon\dot{c} = Fc\,,$$

where the Hermitian matrix F has the elements

$$f_{k\ell} = \langle \varphi_k | W | \varphi_\ell \rangle \qquad (k, \ell \in \mathcal{K}). \tag{3.5}$$

The solution to this problem is thus given by the action of the exponential of F:

$$c(t) = \exp\Bigl(-\frac{it}{\varepsilon}\,F\Bigr)c(0). \tag{3.6}$$

We note that $f_{k\ell} = O(\varepsilon^{3/2})$ if the quadratic Taylor polynomial of W at q vanishes. The computation of the matrix exponential times a vector can then be done efficiently using just a few Lanczos iterations with F, see Sect. III.2.2. The efficient computation of the multi–dimensional integrals in (3.5) is the major computational cost. Sparse Gauss–Hermite quadrature as discussed in Sect. III.1.1 along the eigendirections of the width matrix $\operatorname{Im} PQ^{-1} = (QQ^*)^{-1}$ is a possible computational approach. We refer to Faou, Gradinaru & Lubich (2008) for a discussion of various ways to compute the Galerkin matrix F or its action on a coefficient vector c.

Abstract Formulation of the Time-Stepping Algorithm. For given Gaussian parameters $\Gamma^0 = (q^0, p^0, Q^0, P^0, S^0)$ and coefficients $c^0 = (c_k^0)_{k\in\mathcal{K}}$ as initial data, we denote

- by $\mathcal{T}_t(\Gamma^0, c^0)$ the solution to the free Schrödinger equation given by (3.3),
- by $\mathcal{U}_t(\Gamma^0, c^0)$ the solution of the quadratic-potential equation given by (3.4),
- and by $\mathcal{W}_t(\Gamma^0, c^0)$ the propagator given by (3.6).

With both propagators $\mathcal{U}_t$ and $\mathcal{W}_t$, the parameters q and Q remain constant. Moreover, the propagators $\mathcal{U}_t$ and $\mathcal{W}_t$ commute. This is seen from the expressions in (3.4) and (3.6), noting that only q and Q, but not p and P, enter into the definition of the matrix F.

For a given step size Δt, the time-stepping algorithm is described briefly as follows:

1. *Half-step of the kinetic part.* We define the parameters $(\Gamma^{1/2,-}, c^0)$ by applying the propagator $\mathcal{T}_{\Delta t/2}$ starting from (Γ^0, c^0). This yields updates $q^{1/2}$, $Q^{1/2}$ and $S^{1/2,-}$.
2. *Full step of the potential part.* We split the potential V into its quadratic Taylor polynomial $U^{1/2}$ at $q^{1/2}$ and the remainder term $W^{1/2}$:
$$V(x) = U^{1/2}(x) + W^{1/2}(x)\,.$$
   - We determine the parameters $(\Gamma^{1/2,+}, c^0)$ by applying the propagator $\mathcal{U}_{\Delta t}$ associated with the quadratic potential $U^{1/2}$ starting from $(\Gamma^{1/2,-}, c^0)$. This yields updates p^1, P^1 and $S^{1/2,+}$.
   - We determine the coefficients c^1 using the propagator $\mathcal{W}_{\Delta t}$ associated with the non-quadratic remainder $W^{1/2}$ starting from c^0.
3. *Half-step of the kinetic part.* We define the parameters (Γ^1, c^1) by applying the propagator $\mathcal{T}_{\Delta t/2}$ starting from $(\Gamma^{1/2,+}, c^1)$. This yields updates q^1, Q^1 and S^1.

The Practical Time-Stepping Algorithm. We now give a full algorithmic description. Assume that the stepsize Δt is given, and let the real d-vectors q^n, p^n, the complex $d\times d$ matrices Q^n, P^n, the real scalar S^n, and the complex coefficient vector $c^n = \left(c_k^n\right)_{k\in\mathcal{K}}$ be such that
$$\psi^n = e^{iS^n/\varepsilon} \sum_{k\in\mathcal{K}} c_k^n \,\varphi_k^\varepsilon[q^n, p^n, Q^n, P^n]$$
is an approximation to the solution of the Schrödinger equation (0.1) at time t^n. To compute the approximation ψ^{n+1} at time $t^{n+1} = t^n + \Delta t$ we proceed as follows:

1. Compute $q^{n+1/2}$, $Q^{n+1/2}$, and $S^{n+1/2,-}$ via
$$\begin{aligned} q^{n+1/2} &= q^n + \frac{\Delta t}{2m}\, p^n \\ Q^{n+1/2} &= Q^n + \frac{\Delta t}{2m}\, P^n \\ S^{n+1/2,-} &= S^n + \frac{\Delta t}{4m}\, |p^n|^2\,. \end{aligned} \tag{3.7}$$

2. Compute p^{n+1}, P^{n+1}, and $S^{n+1/2,+}$ via

$$\begin{aligned} p^{n+1} &= p^n - \Delta t\,\nabla V(q^{n+1/2}) \\ P^{n+1} &= P^n - \Delta t\,\nabla^2 V(q^{n+1/2})Q^{n+1/2} \\ S^{n+1/2,+} &= S^{n+1/2,-} - \Delta t\,V(q^{n+1/2})\,. \end{aligned} \tag{3.8}$$

3. Update the coefficient vector $c^{n+1} = (c_k^{n+1})_{k\in\mathcal{K}}$ as

$$c^{n+1} = \exp\Bigl(-\frac{i\Delta t}{\varepsilon}F^{n+1/2}\Bigr)\,c^n. \tag{3.9}$$

Here, $F^{n+1/2} = (f_{k\ell})_{k,\ell\in\mathcal{K}}$ is the Hermitian matrix with entries

$$f_{k\ell} = \langle \varphi_k^{n+1/2}\,|\,W^{n+1/2}\,|\,\varphi_\ell^{n+1/2}\rangle, \tag{3.10}$$

where $\varphi_k^{n+1/2} = \varphi_k^\varepsilon[q^{n+1/2}, p^{n+1}, Q^{n+1/2}, P^{n+1}]$ are the Hagedorn basis functions and

$$W^{n+1/2}(x) = V(x) - U^{n+1/2}(x)$$

is the remainder in the local quadratic approximation to V, given at $q = q^{n+1/2}$ by $U^{n+1/2}(x) = V(q) + \nabla V(q)\,(x-q) + \frac{1}{2}(x-q)^T\nabla^2 V(q)\,(x-q)$.

4. Compute q^{n+1}, Q^{n+1}, and S^{n+1} via

$$\begin{aligned} q^{n+1} &= q^{n+1/2} + \frac{\Delta t}{2m}\,p^{n+1} \\ Q^{n+1} &= Q^{n+1/2} + \frac{\Delta t}{2m}\,P^{n+1} \\ S^{n+1} &= S^{n+1/2,+} + \frac{\Delta t}{4m}\,|p^{n+1}|^2\,. \end{aligned} \tag{3.11}$$

Clearly, Step 3. treating the non-quadratic part of the potential is the computationally most expensive part of the algorithm, since it requires the computation of the multi-dimensional integrals (3.10) and the evaluation of (3.9). The latter can be done efficiently by a few Lanczos iterations as studied in Sect. III.2.2. For the computation of the matrix elements of $F^{n+1/2}$ we refer to the discussion in Faou, Gradinaru & Lubich (2008).

The matrix $F^{n+1/2}$ depends only on $q^{n+1/2}$ and $Q^{n+1/2}$, but not on p^{n+1} and P^{n+1}, since the imaginary parts in the arguments of the Gaussian cancel out in (3.10), and $\operatorname{Im}(P^{n+1}Q^{n+1})^{-1} = (Q^{n+1}Q^{n+1^*})^{-1}$ is independent of P^{n+1} by (1.2).

Properties of the Algorithm. The algorithm is of second-order accuracy in the parameters q, p, Q, P, S and c_k. It enjoys a number of attractive conservation and limit properties:

1. By construction, the algorithm is time-reversible. This holds true because the propagators $\mathcal{U}_t$ and $\mathcal{W}_t$ commute, and in the steps with the potential, the position q and the width matrix $\operatorname{Im} PQ^{-1} = (QQ^*)^{-1}$ remain unchanged.

2. The algorithm preserves the symplecticity relations (1.1) between the matrices Q and P, since it is a composition of exact flows with no or a quadratic potential, and Q and P are not modified in the step with the non-quadratic remainder.
3. The algorithm preserves the L^2 norm of the wave packet, since the Hagedorn functions φ_k are orthonormal and the propagation of the coefficients (c_k) is unitary.
4. For the position and momentum parameters q and p, the algorithm coincides with the Störmer–Verlet method (I.3.5) for the corresponding classical equations of motion.
5. In the limit of taking the full basis set φ_k with all $k \in \mathbb{N}^d$, the Galerkin approximation used in the remainder propagator $\mathcal{W}_t$ becomes exact. Since $\mathcal{U}_t$ and $\mathcal{W}_t$ commute, the second step in the above algorithm becomes the solution of equation (3.2) in this limit. Hence, the algorithm then becomes the Strang splitting $\exp(-\frac{i}{\varepsilon}\Delta t H) \approx \exp(-\frac{i}{\varepsilon}\frac{\Delta t}{2}T)\exp(-\frac{i}{\varepsilon}\Delta t V)\exp(-\frac{i}{\varepsilon}\frac{\Delta t}{2}T)$ of the time-dependent Schrödinger equation.
6. The algorithm is robust in the classical limit $\varepsilon \to 0$: the propagator $\mathcal{W}_{\Delta t}$ of the remainder is $\mathcal{O}(\varepsilon^{1/2}\Delta t)$ close to the identity operator, since $W^{1/2}(x)$ is at least cubic in $(x - q^{1/2})$, and hence the approximation in the potential part becomes exact for $\varepsilon \to 0$. The kinetic part is solved exactly for all ε.

With the exception of Property 5, the above properties are valid also with an approximate computation of the integrals in (3.5). A detailed error analysis in dependence of the time-step Δt, the multi-index set $\mathcal{K}$, and the semi-classical parameter ε is currently under investigation. Here we illustrate some aspects of the method by a numerical experiment as in Faou, Gradinaru & Lubich (2008).

Example 3.1. We consider the time-dependent Schrödinger equation (0.1) on $\mathbb{R}^2$ with $\varepsilon = 10^{-2}$ and a Henon-Heiles potential $V(x_1, x_2) = \frac{1}{2}(x_1^2 + x_2)^2 + 2x_1x_2^2 - \frac{1}{6}x_1^3$. As initial data we choose a Gaussian (1.4) with $q^0 = (-0.1, 0)$, $p^0 = (0, 0.1)$ and $Q^0 = I$, $P^0 = iI$. We applied the above method with step size $\Delta t = 0.1$ and coefficients c_k for $k = (k_1, k_2) \in \{0, \dots, 7\} \times \{0, \dots, 7\}$. The integrals (3.10) were computed approximately using tensor-product Gauss-Hermite quadrature with 8 points in each coordinate direction after the (time-dependent) coordinate transformation $\widehat{x} = B^{1/2}(x - q)$ with the symmetric positive definite matrix $B = \operatorname{Im}(PQ^{-1}) = (QQ^*)^{-1}$. Figure 3.1 shows the values of the squared absolute value of the approximate wave function in the quadrature points (with bilinear interpolation between the points) at times $t = 10$ and 12. The quadrature points adapt to the time-dependent Gaussian and thus form the nodes of a *flying carpet*. The potential and the Hagedorn functions are evaluated only at these quadrature points at every time step.

Remark 3.2. The above approach allows for a simple mixing of classical and quantum modelling: choosing just one quadrature point in a coordinate direction corresponds to evaluating the potential only on a single classical trajectory along that coordinate, whereas capturing fuller quantum behaviour along a coordinate requires more quadrature points.

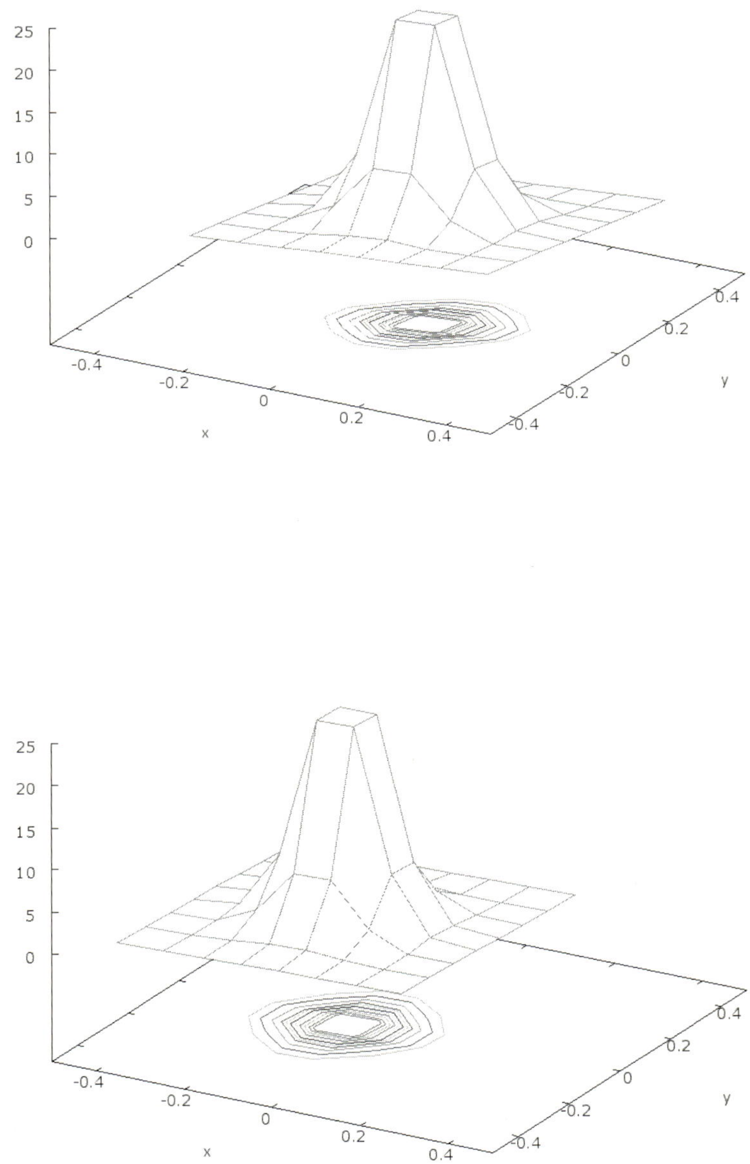

Fig. 3.1. Squared absolute values of the approximate wave function evaluated on the flying carpet of quadrature points, t=10 and 12.

Bibliography

M. Abramowitz and I.A. Stegun, *Handbook of Mathematical Functions*, Dover, New York, 1965. *[III.1], [III.2]*

M. Baer & G.D. Billing, eds., *The Role of Degenerate States in Chemistry*, Advances in Chemical Physics, Vol. 124, Wiley, 2002. *[II.1], [II.2]*

M.H. Beck, A. Jäckle, G.A. Worth & H.-D. Meyer, *The multiconfiguration time-dependent Hartree (MCTDH) method: A highly efficient algorithm for propagating wavepackets*, Phys. Reports 324 (2000), 1–105. *[II.3], [IV.3]*

M.H. Beck & H.-D. Meyer, *An efficient and robust integration scheme for the equations of motion of the multiconfiguration time-dependent Hartree (MCTDH) method*, Z. Phys. D 42 (1997), 113–129. *[IV.3]*

M. Ben-Nun & T.J. Martinez, *Nonadiabatic molecular dynamics: validation of the multiple spawning method for a multidimensional problem*, J. Chem. Phys. 108 (1998), 7244–7257. *[II.5]*

M. Ben-Nun & T.J. Martinez, *A multiple spawning approach to tunneling dynamics*, J. Chem. Phys. 112 (2000), 6113–6121. *[II.5]*

M.V. Berry, *Quantal phase factors accompanying adiabatic changes*, Proc, Royal Soc. London. Series A, Math. Phys. Sci. 392 (1984), 45–57. *[II.2]*

G.D. Billing, *The Quantum Classical Theory*, Oxford Univ. Press, 2003. *[II.5]*

S. Blanes, F. Casas & A. Murua, *Symplectic splitting operator methods for the time-dependent Schrödinger equation*, J. Chem. Phys. 124, 234105 (2006) (10 pages). *[III.2]*

S. Blanes, F. Casas & J. Ros, *Improved high order integrators based on the Magnus expansion*, BIT 40 (2000), 434–450. *[III.4]*

S. Blanes & P.C. Moan, *Splitting methods for the time-dependent Schrödinger equation*, Physics Letters A 265 (2000), 35–42. *[III.4]*

S. Blanes & P.C. Moan, *Practical symplectic partitioned Runge–Kutta and Runge–Kutta–Nyström methods*, J. Comput. Appl. Math. 142 (2002), 313–330. *[III.3]*

S. Brandt & H.D. Dahmen, *The Picture Book of Quantum Mechanics*, 3rd ed., Springer, New York, 2001. *[I.0]*

M. Born, *Quantenmechanik der Stoßvorgänge*, Z. Phys. 38 (1926), 803–827. *[I.1]*

M. Born & V. Fock, *Beweis des Adiabatensatzes*, Z. Phys. 51 (1928), 165–180. *[II.2]*

F. A. Bornemann and C. Schütte, *On the singular limit of the quantum-classical molecular dynamics model*, SIAM J. Appl. Math. 59 (1999), 1208–1224. *[II.5]*

L. de Broglie, *Recherche sur la théorie des quanta*, Thèse, Paris, 1924. http://tel.archives-ouvertes.fr/tel-00006807 *[I.2]*

I. Burghardt, H.-D. Meyer, & L. S. Cederbaum. *Approaches to the approximate treatment of complex molecular systems by the multiconfiguration time-dependent Hartree method*, J. Chem. Phys. 111 (1999), 2927–2939. *[II.3], [II.5]*

M. Caliari, M. Vianello & L. Bergamaschi, *Interpolating discrete advection-diffusion propagators at Leja sequences*, J. Comput. Appl. Math. 172 (2004), 79–99. *[III.2]*

E. Cancès & C. Le Bris, *On the time-dependent Hartree–Fock equations coupled with a classical nuclear dynamics*, Math. Mod. Meth. Appl. Sci 9 (1999), 963–990. *[II.5]*

C. Canuto, A. Quarteroni, M.Y. Hussaini, T.A. Zang, *Spectral Methods: Fundamentals in Single Domains*, Springer, Berlin, 2006. *[III.1]*

J.M. Chadam & R.T. Glassey, *Global existence of solutions to the Cauchy problem for time-dependent Hartree equations*, J. Math. Phys. 16 (1975), 1122–1130. *[II.3]*

P.G. Ciarlet, *Introduction to the Finite Element Method*, in Handbook of Numerical Analysis, Vol. II (P.G. Ciarlet and J.L. Lions, eds.), North-Holland, Amsterdam, 1991, 17–352. *[II.6]*

C.W. Clenshaw, *Chebyshev Series for Mathematical Functions*, Mathematical Tables, Vol. 5, H.M. Stationery Office, London, 1962. *[III.2]*

R.D. Coalson & M. Karplus, *Multidimensional variational Gaussian wavepackets dynamics with application to photodissociation spectroscopy*, J. Chem. Phys. 93 (1990), 3919–3930. *[II.4]*

C. Cohen-Tannoudji, B. Diu & F. Laloë, *Quantum Mechanics*, Vols. I & II, Wiley, New York, 1977. *[I.0]*

P.A.M. Dirac, *Note on exchange phenomena in the Thomas atom*, Proc. Cambridge Phil. Soc. 26 (1930), 376–385. *[.0]*

P.A.M. Dirac, *The Principles of Quantum Mechanics*, Clarendon Press, Oxford, 1930. *[I.0]*

W. Domcke, D.R. Yarkony & H. Köppel (eds.), *Conical Intersections: Electronic Structure, Dynamics & Spectroscopy*, World Scientific, 2004. *[II.2]*

V.L. Druskin & L.A. Knizhnerman, *Krylov subspace approximations of eigenpairs and matrix functions in exact and computer arithmetic*, Numer. Lin. Alg. Appl. 2 (1995), 205 – 217. *[III.2]*

P. Ehrenfest, *Bemerkung über die angenäherte Gültigkeit der klassischen Mechanik innerhalb der Quantenmechanik*, Z. Phys. 45 (1927), 455–457. *[I.4]*

A. Einstein, *Über einen die Erzeugung und Verwandlung des Lichtes betreffenden heuristischen Gesichtspunkt*, Ann. Phys. 17 (1905), 132–148. *[I.2]*

E. Faou & V. Gradinaru, *Gauss-Hermite wavepacket dynamics: convergence of the spectral and pseudo-spectral approximation*, Report, Math. Inst., Univ. Tübingen, 2007. *[III.1]*

E. Faou, V. Gradinaru & C. Lubich, *Computing semi-classical quantum dynamics with Hagedorn wave packets*, Preprint, Math. Inst., Univ. Tübingen, 2008. *[V.3]*

E. Faou & C. Lubich, *A Poisson integrator for Gaussian wavepacket dynamics*, Comput. Vis. Sci. 9 (2006), 45–55. *[II.4], [IV.4]*

M.D. Feit, J.A. Fleck Jr. & A. Steiger, *Solution of the Schrödinger equation by a spectral method*, J. Comput. Phys. 47 (1982), 412–433. *[III.3]*

J.A. Fleck Jr., J.R. Morris & M.D. Feit, *Time-dependent propagation of high energy laser beams through the atmosphere*, Appl.Phys. A: Materials Science & Processing 10 (1976), 129–160. *[III.3]*

H. Feldmeier & J. Schnack, *Molecular dynamics for fermions*, Rev. Modern Phys. 72 (2000), 655–688. *[II.1]*

B. Fornberg, *A Practical Guide to Pseudospectral Methods*, Cambridge Univ. Press, 1996. *[III.1]*

J. Frenkel, *Wave Mechanics, Advanced General Theory*, Clarendon Press, Oxford, 1934. *[II.1]*

W. Gautschi, *Numerical Analysis, An Introduction*, Birkhäuser, Basel, 1997. *[III.1]*

T. Gerstner & M. Griebel, *Numerical integration using sparse grids*, Numer. Algor. 18 (1998), 209–232. *[III.1]*

G.H. Golub & C.F. Van Loan, *Matrix Computations*, Third edition, The Johns Hopkins University Press, Baltimore, 1996. *[III.2]*

D. Gottlieb & S.A. Orszag, *Numerical Analysis of Spectral Methods: Theory and Applications*, SIAM, Philadelphia, 1977. *[III.1]*

V. Gradinaru, *Fourier transform on sparse grids: code design and application to the time dependent Schrödinger equation*, Computing 80 (2007), 1–22. *[III.1]*

V. Gradinaru, *Strang splitting for the time-dependent Schrödinger equation on sparse grids*, SIAM J. Numer. Anal. 46 (2007), 103–123. *[III.1], [V.0]*

S.K. Gray & D.E. Manolopoulos, *Symplectic integrators tailored to the time-dependent Schrödinger equation*, J. Chem. Phys. 104 (1996), 7099–7112. *[III.2]*

S.J. Gustafson & I.M. Sigal, *Mathematical Concepts of Quantum Mechanics*, Springer, Berlin, 2003. *[I.0], [I.3]*

G.A. Hagedorn, *Semiclassical quantum mechanics. I. The $\hbar \to 0$ limit for coherent states*, Comm. Math. Phys. 71 (1980), 77–93. *[II.4], [V.1]*

G.A. Hagedorn, *Raising and lowering operators for semiclassical wave packets*, Ann. Phys. 269 (1998), 77–104. *[V.0], [V.1], [V.2]*

E. Hairer, C. Lubich & G. Wanner, *Geometric numerical integration illustrated by the Störmer–Verlet method*, Acta Numerica 12 (2003), 399–450. *[I.3]*

E. Hairer, C. Lubich & G. Wanner, *Geometric Numerical Integration: Structure-Preserving Algorithms for Ordinary Differential Equations*, Second edition. Springer, Berlin, 2006. *[II.4], [III.3], [IV.2], [IV.3]*

E. Hairer, S.P. Nørsett & G. Wanner, *Solving Ordinary Differential Equations. I. Nonstiff Problems*, Second edition. Springer, Berlin, 1993. *[IV.2]*

K. Hallatschek, *Fouriertransformation auf dünnen Gittern mit hierarchischen Basen*, Numer. Math. 63 (1992), 83–97. *[III.1]*

E. Hansen & A. Ostermann, *Exponential splitting for unbounded operators*, Report, Inst. für Mathematik, Univ. Innsbruck, 2008. *[III.3]*

R.H. Hardin & F.D. Tappert, *Applications of the split-step Fourier method to the numerical solution of nonlinear and variable coefficient wave equations*, SIAM Rev. 15 (1973), 423. *[III.3]*

W. Heisenberg, *Über quantentheoretische Umdeutung kinematischer und mechanischer Beziehungen*, Z. Phys. 33 (1925), 879–893. *[I.4]*

W. Heisenberg, *Über den anschaulichen Inhalt der quantentheoretischen Kinematik und Mechanik*, Z. Phys. 43 (1927), 172–198. *[I.4]*

E.J. Heller, *Time dependent variational approach to semiclassical dynamics*, J. Chem. Phys. 64 (1976), 63–73. *[II.1], [II.4]*

E.J. Heller, *Frozen Gaussians: a very simple semiclassical approximation*, J. Chem. Phys. 75 (1981), 2923–2931. *[II.4], [II.5]*

P.D. Hislop & I.M. Sigal, *Introduction to Spectral Theory. With Applications to Schrödinger Operators*, Springer, Berlin, 1996. *[I.3]*

M. Hochbruck & C. Lubich, *On Krylov subspace approximations to the matrix exponential operator*, SIAM J. Numer. Anal. 34 (1997), 1911–1925. *[III.2]*

M. Hochbruck & C. Lubich, *On Magnus integrators for time-dependent Schrödinger equations*, SIAM J. Numer. Anal. 41 (2003), 945–963. *[III.4]*

M. Hochbruck, C. Lubich & H. Selhofer, *Exponential integrators for large systems of differential equations*, SIAM J. Sci. Comput. 19 (1998), 1552–1574. *[III.2]*

W. Hundsdorfer & J. Verwer, *Numerical Solution of Time-Dependent Advection-Diffusion-Reaction Equations*, Springer, Berlin, 2003. *[IV.2]*

A. Iserles, H.Z. Munthe-Kaas, S.P. Nørsett & A. Zanna, *Lie-group methods*, Acta Numerica (2000), 215–365. *[III.4]*

A. Iserles & S.P. Nørsett, *On the solution of linear differential equations in Lie groups*, R. Soc. Lond. Philos. Trans. Ser. A Math. Phys. Eng. Sci. 357 (1999), 983–1019. *[III.4]*

T. Jahnke & C. Lubich, *Error bounds for exponential operator splittings*, BIT 40 (2000), 735–744. *[III.3]*

T. Jahnke & W. Huisinga, *A dynamical low-rank approach to the chemical master equation*, Report, FU Berlin, 2007. *[II.3]*

W. Kahan & R.-C. Li, *Composition constants for raising the orders of unconventional schemes for ordinary differential equations*, Math. Comput. 66 (1997), 1089–1099. *[III.3]*

T. Kato, *On the adiabatic theorem of quantum mechanics*, Phys. Soc. Jap. 5 (1950), 435–439. *[II.2]*

T. Kato, *Fundamental properties of Hamiltonian operators of Schrödinger type*, Trans. Amer. Math. Soc. 70 (1951), 195–211. *[I.5]*

T. Kato, *Perturbation Theory for Linear Operators*, 2nd ed., Springer, Berlin, 1980. *[I.3]*

Y. Katznelson, *An Introduction to Harmonic Analysis*, Dover Publ., New York, 1976. *[I.2]*

A.K. Kerman & S.E. Koonin, *Hamiltonian formulation of time-dependent variational principles for the many-body system*, Ann. Phys. 100 (1976), 332–358. *[II.1]*

O. Koch, W. Kreuzer & A. Scrinzi. *Approximation of the time-dependent electronic Schrödinger equation by MCTDHF*, Appl. Math. Comput. 173 (2006), 960–976. *[II.3]*

O. Koch & C. Lubich, *Regularity of the multi-configuration time-dependent Hartree approximation in quantum molecular dynamics*, M2AN 41 (2007), 315–332. *[II.3]*

O. Koch & C. Lubich, *Dynamical low rank approximation*, SIAM J. Matrix Anal. Appl. 29 (2007), 434–454. *[II.3]*

P. Kramer & M. Saraceno, *Geometry of the Time-Dependent Variational Principle in Quantum Mechanics*, Lecture Notes in Physics 140, Springer, Berlin, 1981. *[II.1]*

C. Lanczos, *An iteration method for the solution of the eigenvalue problem of linear differential and integral operators*, J. Res. Nat. Bur. Standards 45 (1950), 255–282. *[III.2]*

C. Lasser & S. Teufel, *Propagation through conical crossings: an asymptotic semigroup*, Comm. Pure Appl. Math. 58 (2005), 1188–1230. *[II.2]*

C. Le Bris (ed.), *Handbook of Numerical Analysis, Special Volume Computational Chemistry*, Elsevier, Amsterdam, 2003. *[II.2]*

C. Lubich, *On variational approximations in quantum molecular dynamics*, Math. Comp. 74 (2005), 765–779. *[II.6]*

C. Lubich, *A variational splitting integrator for quantum molecular dynamics*, Appl. Numer. Math. 48 (2004), 355–368. *[IV.2]*

W. Magnus, *On the exponential solution of differential equations for a linear operator*, Comm. Pure Appl. Math. VII (1954), 649–673. *[III.4]*

G. Marchuk, *Some applications of splitting-up methods to the solution of mathematical physics problems*, Aplikace Matematiky 13 (1968), 103–132. *[III.3]*

P.A. Markowich, P. Pietra, C. Pohl, *Numerical approximation of quadratic observables of Schrödinger-type equations in the semi-classical limit*. Numer. Math. 81 (1999), 595–630.

A.I. Markushevich, *Theory of Functions of a Complex Variable*, Chelsea Publ, Comp., New York, 1977. *[III.2]*

J.E. Marsden & T.S. Ratiu, *Introduction to Mechanics and Symmetry. A Basic Exposition of Classical Mechanical Systems*, Second edition, Texts in Applied Mathematics 17, Springer-Verlag, New York, 1999. *[II.1], [II.4]*

A.D. McLachlan, *A variational solution of the time-dependent Schrodinger equation*, Mol. Phys. 8 (1964), 39–44. *[II.1]*

R.I. McLachlan, *On the numerical integration of ordinary differential equations by symmetric composition methods*, SIAM J. Sci. Comput. 16 (1995), 151–168. *[III.3]*

R.I. McLachlan & G.R.W. Quispel, *Splitting methods*, Acta Numerica 11 (2002), 341–434. *[III.3]*

A. Messiah, *Quantum Mechanics*, Dover Publ., New York, 1999 (reprint of the two-volume edition published by Wiley, 1961–1962). *[I.0]*

H.-D. Meyer, U. Manthe & L. S. Cederbaum, *The multi-configurational time-dependent Hartree approach*, Chem. Phys. Lett. 165 (1990), 73–78. *[II.3]*

C. Moler & C. Van Loan, *Nineteen dubious ways to compute the exponential of a matrix, twenty-five years later*, SIAM Rev. 45 (2003), 3–49. *[III.2]*

J.J. Monaghan, *Smoothed particle hydrodynamics*, Annu. Rev. Astron. Astrophys. 30 (1992), 543–574. *[II.5]*

J. von Neumann, *Mathematische Grundlagen der Quantenmechanik*, Springer, Berlin, 1932, Reprint 1996. *[I.0], [I.3]*

A. Nonnenmacher & C. Lubich, *Dynamical low-rank approximation: applications and numerical experiments*, Math. Comput. Simulation, 2008, to appear. *[II.3]*

T.J. Park & J.C. Light, *Unitary quantum time evolution by iterative Lanczos reduction*, J. Chem. Phys. 85 (1986), 5870–5876. *[III.2]*

A. Raab, G.A. Worth, H.-D. Meyer & L.S. Cederbaum, *Molecular dynamics of pyrazine after excitation to the S_2 electronic state using a realistic 24-mode model Hamiltonian*, J. Chem. Phys. 110 (1999), 936–946. *[II.3]*

M. Reed & B. Simon, *Methods of Modern Mathematical Physics. I: Functional Analysis*, Academic Press, New York, 1972. *[I.3]*

M. Reed & B. Simon, *Methods of Modern Mathematical Physics. II: Fourier Analysis, Self-Adjointness*, Academic Press, New York, 1975. *[I.2], [I.3], [I.5]*

T.J. Rivlin, *Chebyshev Polynomials*, Second edition, Wiley, New York, 1990. *[III.2]*

D.J. Rowe, A. Ryman & G. Rosensteel, *Many-body quantum mechanics as a symplectic dynamical system*, Phys. Rev. A 22 (1980), 2362–2373. *[II.1]*

Y. Saad, *Analysis of Some Krylov Subspace Approximations to the Matrix Exponential Operator*, SIAM J. Numer. Anal. 29 (1992), 209–228. *[III.2]*

E. Schrödinger, *Quantisierung als Eigenwertproblem (Vierte Mitteilung)*, Ann. Phys. 81 (1926), 109–139. *[I.1]*

B. Simon, *Holonomy, the quantum adiabatic theorem, and Berry's phase*, Phys. Rev. Letters 51 (1983), 2167–2170. *[II.2]*

S.A. Smolyak, *Quadrature and interpolation formulas for tensor products of certain classes of functions*, Dokl. Akad. Nauk SSSR 4 (1963), 240–243. *[III.1]*

M. Sofroniou & G. Spaletta, *Derivation of symmetric composition constants for symmetric integrators*, J. of Optimization Methods and Software (2005), 597 – 613. *[III.3]*

M.H. Stone, *On One-Parameter Unitary Groups in Hilbert Space*, Ann. Math., 2nd Ser., 33 (1932), 643–648. *[I.3]*

G. Strang, *On the construction and comparison of difference schemes*, SIAM J. Numer. Anal. 5 (1968), 506–517. *[III.3]*

M. Suzuki, *Fractal decomposition of exponential operators with applications to many-body theories and Monte Carlo simulations*, Phys. Lett. A 146 (1990) 319–323. *[III.3]*

A. Szabo & N.S. Ostlund, *Modern Quantum Chemistry: Introduction to Advanced Electronic Structure Theory*, Revised ed., Dover Publ., New York, 1996. *[II.2], [II.3]*

H. Tal-Ezer & R. Kosloff, *An accurate and efficient scheme for propagating the time dependent Schrödinger equation*, J. Chem. Phys. 81 (1984), 3967–3971. *[III.2]*

D.J. Tannor, *Introduction to Quantum Mechanics: A Time-Dependent Perspective*, University Science Books, Sausalito, 2007. *[I.0]*

S. Teufel, *Adiabatic Perturbation Theory in Quantum Dynamics*, Lecture Notes in Mathematics, 1821. Springer-Verlag, Berlin, 2003. *[II.2]*

M. Thalhammer, *High-order exponential operator splitting methods for time-dependent Schrödinger equations*, SIAM J. Numer. Anal. 46 (2008), 2022–2038.

B. Thaller, *Visual Quantum Mechanics*, Springer, New York, 2000. *[I.0], [III.1]*

L.N. Trefethen, *Spectral Methods in MATLAB*, SIAM, Philadelphia, 2000. *[III.1]*

L.N. Trefethen & D. Bau, III, *Numerical Linear Algebra*, SIAM, Philadelphia, 1997. *[III.2]*

H.F. Trotter, *On the product of semigroups of operators*, Proc. Amer. Math. Soc. 10 (1959), 545–551. *[I.3]*

H. Wang & M. Thoss, *Multilayer formulation of the multiconfiguration time-dependent Hartree theory*, J. Chem. Phys. 119 (2003), 1289–1299. *[II.3]*

H. Yoshida, *Construction of higher order symplectic integrators*, Phys. Lett. A 150 (1990), 262–268. *[III.3]*

H. Yserentant, *A new class of particle methods*, Numer. Math. 76 (1997), 87–109. *[II.5]*

J. Zanghellini, M. Kitzler, C. Fabian, T. Brabec & A. Scrinzi, *An MCTDHF approach to multielectron dynamics in laser fields*, Laser Phys. 13 (2003), 1064–1068. *[II.3]*

C. Zenger, *Sparse grids*, in W. Hackbusch (ed.), Parallel algorithms for partial differential equations, Notes on Numerical Fluid Mechanics 31, Vieweg, Braunschweig, 1991, 241-251. *[III.1]*

Index